Singapore Math®
Tests

5A

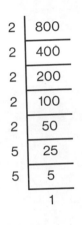

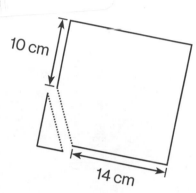

10 cm

14 cm

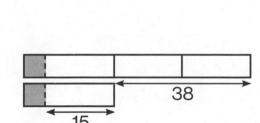

38

15

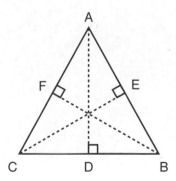

$$28 - 2 \times \{[3 \times (6 \div 2 + 7) + 5] \div 7\} =$$

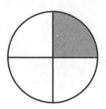
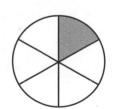

Differentiated Unit Tests
Continual Assessments

 Singapore Math Inc.®

BLANK

Preface

Singapore Math® Tests is a series of structured assessments to help teachers evaluate student progress. The tests align with the content of Primary Mathematics Common Core textbooks.

Each level offers differentiated tests (Test A and Test B) to suit individual needs. Tests consist of multiple-choice questions that assess comprehension of key concepts and free response questions that demonstrate problem solving skills. Three continual assessments cover topics from earlier units and a year-end assessment covers the entire curriculum.

Test A focuses on key concepts and fundamental problem solving skills.

Test B focuses on the application of analytical skills, thinking skills, and heuristics.

Contents

BLANK

Name: _____   Date: _____

Test A   30 min

Score

Unit 1 Whole Numbers

Section A (2 points each)
Circle the correct option: **A**, **B**, **C**, or **D**.

1. What is the place value of the digit 7 in the number below?

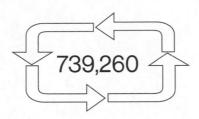

739,260

A hundred thousands   **B** ten thousands

C thousands   **D** hundreds

2. Alberto drove from New York to Los Angeles. He drove a total of
    4,631,588 m. Round this distance to the nearest 1,000 m.

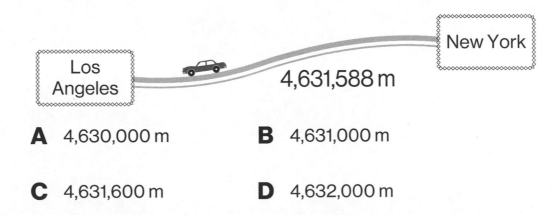

A 4,630,000 m   **B** 4,631,000 m

C 4,631,600 m   **D** 4,632,000 m

3. Which one of the following gives a value of 10^5?

    A 50,000 × 20   **B** 1,250 × 80

    C 25 × 400   **D** 6,000,000 ÷ 6

4. What is the value of $9^2 \times 10^3$?

    A 90   **B** 8,100

    C 9,000   **D** 81,000

5. Which of the following shows the prime factorization of 48?

    A $2^4 \times 3^2$ **B** $2^3 \times 3^2$

    C $2^4 \times 3$ **D** $2^2 \times 3^3$

Section B (2 points each)

6.   Write 5,043,708 in words.

7.   Write 982 million, 653 thousand, 7 ones in expanded form.

8.   Write the missing number.

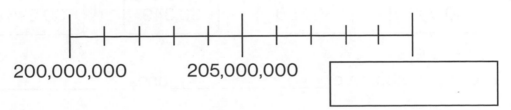

200,000,000   205,000,000

9. Fill in the blank.

300,000 less than 2,843,090 is _____.

10. Fill in the blanks with the correct answers.

In 1,439,580,

the value of the digit 9 is _____.

the digit 3 is in the _____ place.

the digit 1 stands for _____.

the digit _____ is in the hundred thousands place.

the value of the digit 3 is _____ more than the value of the digit 9.

11. Two of the following numbers when rounded to the nearest hundred are 30,300.

| 30,249 | 30,350 | 30,349 | 30,295 |

The two numbers are _____ and _____.

12. Arrange these numbers in increasing order.

| 2,017,380 | 2,701,038 | 2,107,308 |

_____ , _____ , _____

13. Cross out (x) the numbers that are **not** factors of the given numbers.

a) Factors of 6:

| 1 | 2 | 3 | 4 | 5 | 6 |

b) Factors of 15:

| 1 | 3 | 5 | 8 | 10 | 15 |

14. What is the lowest common multiple of 6 and 9?

15. What is the greatest common factor of 20 and 45?

Section C (4 points each)

16.   A total of $1,823,450 was raised at a charity event. What was the amount when rounded to the nearest thousand?

17.   Thomas has less than 50 m of wire. He can cut the wire into equal pieces of length 3 m or 7 m without any wire left over. What is the longest possible length of wire Thomas has?

18. A factory packed 4,820 cans of soup into 20 boxes. Round the number of cans to the nearest thousand and estimate the number of cans in each box.

19. There are 500 sheets of paper in a box. How many sheets of paper are there in 888 such boxes?

Tests 5A

20. A shop owner paid $70,000 for 5,000 soccer uniforms to sell in his store. What was the cost of each soccer uniform?

BLANK

Name: _____ Date: _____

Test B

30 min

50

Score

Unit 1 Whole Numbers

Section A (2 points each)
Circle the correct option: **A**, **B**, **C**, or **D**.

1. Given that 39 × 68 = 2,652, what is the value of the expression below?

2,652,000 ÷ 3,900

A 68

B 680

C 6,800

D 68,000

2. Which of the following numbers has the digit 4 in its hundred thousands place?

A 4,735,610

B 2,470,538

C 649,705

D 1,384,705

3. What is 10 to the seventh power?

   A 50,000   **B** 1,000,000

   C 10,000,000   **D** 5,000,000

4. Which of the following shows the prime factorization of 36?

   A $2^2 \times 3^2$ **B** $2^3 \times 3^6$

   C $2^3 \times 3$ **D** 3×6

5. Which of the following lists both the greatest common factor and the lowest common multiple of 8 and 12?

   A 1, 24   **B** 4, 24

   C 4, 8   **D** 2, 24

Section B (2 points each)

6. Find the value of $528,000 \div 10^3$.

    [ ]

7. Fill in the blanks and express your answers using exponents.

    a) $27 \times$ _____ $= 27,000$

        $27,000 \div$ _____ $= 27$

    b) $40 \times$ _____ $= 80,000$

        $80,000 \div$ _____ $= 40$

8. What is the lowest common multiple of 3 and 7?

    [ ]

9. In 5,483,629,

the digit _____ is in the hundred thousands place.

the value of the digit 5 is _____.

the digit 3 stands for _____.

the digit 4 is in the _____ place.

the value of the digit 5 is 4,600,000 more than the value of the

digit _____.

10. Form numbers using the given digits. Each digit can be used only once in each answer.

a) greatest 6-digit number _____

b) greatest 6-digit even number _____

c) smallest 6-digit number _____

d) smallest 6-digit odd number _____

11. Arrange the numbers in decreasing order.

| 8,024,135 | 8,024,513 | 8,025,134 | 8,204,315 |

_____ ,

_____ ,

_____ ,

_____

12. Find the prime factorization of 36. Express your answer using exponents.

13. The 3rd multiple of 6 has the same value as the 2nd multiple of

_____.

14. The difference between the greatest factor and the smallest

factor of 57 is _____.

15. Study the pattern carefully.

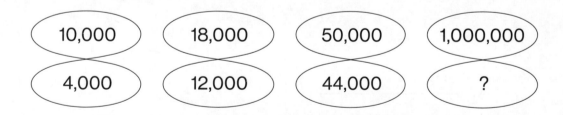

The missing number is _____.

Section C (4 points each)

16. The table below shows the number of visitors to an international airport in two months.

January	4,950,600
February	3,138,500

Round the number of visitors in each month to the nearest hundred thousand. Then estimate the total number of visitors in the two months altogether.

17. Samuel wants to buy a pair of headphones which costs less than $30. The price of the pair of headphones can be exactly divided by both 6 and 9. What is the price of the pair of headphones?

18.	A carpet measures 280 ft by 170 ft. Round the length and width to the nearest hundred feet and estimate the area of the carpet.

19.	Each member of a club pays an annual membership fee of $3,000. There are 614 members. How much does the club collect in membership fees in a year?

20. Errol had $3,450 and his brother had $1,548. They spent a total of $3,920 on a sound system. Round each amount to the nearest $1,000 and estimate the amount of money they had left.

BLANK

Name: _____   Date: _____

Test A   40 min

⬭ **Unit 2** More Calculations with Whole Numbers

50

Score

Section A (2 points each)
Circle the correct option: **A**, **B**, **C**, or **D**.

1.   Find the value of 18 − 12 ÷ (3 × 2) + 2.

   A 14   **B** 6

   C 3   **D** 18

2.   Which of the following gives the same answer as (18 − 3) × 5?

   A (18 × 5) − (3 × 5)   **B** (18 − 5) × (3 − 5)

   C 8 × 5 + 3 × 5   **D** (18 × 5) − 3

3.   Marcus had 52 magnets. After using 34 magnets for a project, he packed the remaining magnets equally into 3 bags. Which one of the following represents the number of magnets in each bag?

   A 52 − 34 ÷ 3   **B** 52 − (34 ÷ 3)

   C (52 − 34) ÷ 3   **D** 34 ÷ 3

4. What is the value of 89 × 22?

   A 1,958   B 1,908

   C 1,780   D 1,948

5. What is the value of 47 divided by 21?

   A 2R3   B 2R6

   C 2R7   D 2R5

Section B (2 points each)

Solve.

6. $30 + 24 \div 4 - 2 = $ ☐

7. $36 \div 9 + 3 \times (8 - 6) = $ ☐

8. $5 \times \{ [(10 - 2) \div 4] + 7 \} = $ ☐

9. Insert parentheses to make the following equation mathematically correct.

$$6 \div 3 \times 2 + 2 = 3$$

10. Fill in the missing numbers.

$(2 + 9) \times 16 = ($ _____ $\times 16) + (9 \times$ _____ $)$

11. Fill in the missing numbers.

$99 \times 5 = (100 \times$ _____ $) - ($ _____ $\times 5)$

12. Multiply.

$$
\begin{array}{r}
4\ 5\ 5 \\
\times\ \ \ 3\ 5 \\
\hline
\\
\hline
\end{array}
$$

13. Divide.

   $4{,}992 \div 16 =$ [   ]

14. The total weight of a cow and a calf is 1,033 lb. The calf weighs 65 lb. What is the weight of the cow?

   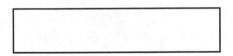

15. The area of a rectangular playground is 4,680 m². What is the length of the playground if the width is 60 m?

   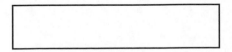

Section C (4 points each)

16. Heleena has $750 and her sister has $390. How much money must Heleena give her sister so that they will each have an equal amount of money?

17. A farmer had 68 pumpkins. He sold half of them for $12 each. How much money did he receive?

18. A baker made 336 muffins and he put 8 in each box. He sold 35 boxes. How many boxes were left unsold?

19. There are 8,454 people at a baseball game. The number of adults at the baseball game is 5 times the number of children. How many adults are there?

20. Pia went shopping with $2,000. She paid $370 for a camera and 4 times as much for a television set. How much money did Pia have left?

Test B

40 min

Unit 2 More Calculations with Whole Numbers

50

Score

Section A (2 points each)
Circle the correct option: **A**, **B**, **C**, or **D**.

1. The sum of two numbers is 58. One number is 14 more than the other number. Which one of the following shows the value of the greater number?

   A $58 + 14 \div 2$

   B $58 - 14 \div 2 + 14$

   C $(58 - 14) \div 2 + 14$

   D $58 + (14 \times 2) \div 2$

2. What is the missing symbol?

   $8 + 12 \boxed{\ ?\ } 6 \div 3 = 18$

   A $+$

   B $-$

   C $\times$

   D $\div$

3. What number when divided by 28 has a quotient of 9 with a remainder of 5?

   A 149

   B 42

   C 257

   D 252

4. What is the value of 1,950 × 27?

    A 5,265   **B** 52,650

    C 51,780   **D** 52,560

5. What is the value of 3,168 divided by 36?

    A 88   **B** 106

    C 880   **D** 36

Section B (2 points each)

6. Fill in the missing numbers.

   $(81 \times 12) + (19 \times 12) = ($ _____ $+$ _____ $) \times 12$

7. Fill in the missing numbers.

   $99{,}997 \times 1{,}000 = (100{,}000 \times$ _____ $) - ($ _____ $\times 1{,}000)$

8. Solve.

   a) $88 + 12 \div 3 - 4 \times 7 =$

   b) $9 \times (14 + 6) \times 2 =$

9. Find the value.

$$28 - 2 \times \{ [3 \times (6 \div 2 + 7) + 5] \div 7 \} = $$

10. Find the value of $99 \times 1{,}001 - 99$.

11. Insert parentheses to make the following equation mathematically correct.

$$4 \times 5 - 3 + 6 \div 2 = 16$$

12. B is a whole number. If you multiply B by 42 and add 1,324 to the result, the answer is 13,588. What number is B?

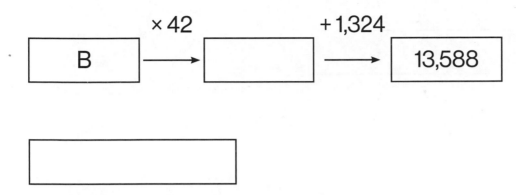

13. Write the missing digit.

$$
\begin{array}{r}
1\ 4\ \boxed{}\ 9 \\
\times \qquad 5 \\
\hline
7\ 1\ 4\ 5
\end{array}
$$

14. Divide 9,930 by 22. What is the answer?

15. 1,536 hours = _____ minutes

Section C (4 points each)

16. A farmer had 2,021 limes. She packed them in bags of 12. How many limes were left unpacked?

17. A sum of money was shared equally between Jess and Mya. After Jess had given $2,100 of her share to Mya, Mya had 5 times as much money as Jess. How much was the sum of money?

18. Two groups of people visited an amusement park. A group of 8 children and 4 adults paid $856 for their tickets. A different group of 2 children and 4 adults paid $508. What was the cost of a ticket for an adult?

19. 6 boxes of candies were sold for $32 in Shop A. 8 such boxes of candies were sold for $36 in Shop B. Sam bought 24 boxes of candies from shop B instead of Shop A. How much money did he save?

20. Andrew collected 1,900 local stamps and 240 foreign stamps. After he received an equal number of local and foreign stamps from his brother, the number of his local stamps was 3 times the number of his foreign stamps. How many foreign stamps did he receive from his brother?

BLANK

Name: _____ Date: _____

Test A   **40 min**

Unit 3 Fractions

Section A (2 points each)
Circle the correct option: **A**, **B**, **C**, or **D**.

1. Which of the following is **not** an equivalent fraction of $\frac{6}{10}$?

    A $\frac{12}{20}$ **B** $\frac{3}{5}$ **C** $\frac{6}{20}$ **D** $\frac{18}{30}$

2. Which of the following is an improper fraction?

    A $\frac{7}{5}$ **B** $\frac{5}{7}$ **C** $1\frac{2}{5}$ **D** 7

3. What is the value of $\frac{1}{2} - \frac{2}{5}$?

    A $\frac{1}{5}$ **B** $\frac{2}{10}$ **C** $\frac{1}{10}$ **D** $\frac{2}{5}$

4. What is the area of a rug measuring 10 yd by $\frac{7}{8}$ yd?

A $8\frac{7}{10}$ yd²   **B** $\frac{35}{8}$ yd²   **C** $8\frac{1}{4}$ yd²   **D** $8\frac{3}{4}$ yd²

5. Luka spent $90 on a bike. This was $\frac{5}{6}$ of his savings. How much savings did he have left?

A $75   **B** $45

C $18   **D** $15

Section B (2 points each)

6. Write the missing numerator.

$$\frac{5}{9} = \frac{\boxed{}}{27}$$

7. Write the missing improper fraction.

$$3\frac{5}{6} - \boxed{} = 2\frac{3}{4}$$

8. Given that $\frac{1}{4} + \frac{\boxed{?}}{8} = 1\frac{1}{2}$, what is the missing numerator?

9. Divide 52 by 6. Give your answer as a mixed number in its simplest form.

10. Express the value of $\frac{3}{8}$ of 9 as a mixed number.

11. Divide 90 by 25 and express the answer as a fraction in its simplest form.

12. Arrange the following from the greatest to the smallest.

$2\dfrac{1}{4}$ 1 $\dfrac{11}{4}$ $1\dfrac{5}{8}$

_____ , _____ , _____ , _____
greatest

13. On the number line below, A and B each stand for a fraction. Find the sum of A and B.

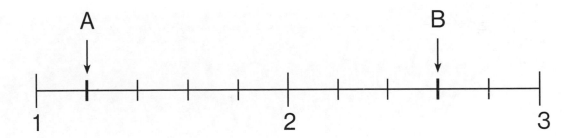

14. Satsuki poured 300 ml of syrup equally into 7 bottles. How much syrup is there is in each bottle?

15. Gabby had $2\frac{1}{3}$ kg of flour. After she used some flour to bake a cake, she had $1\frac{1}{2}$ kg of flour left. How much flour did she use to bake the cake?

Section C (4 points each)

16. A shop owner had 160 cans of juice. She sold $\frac{5}{8}$ of them. How many cans of juice did she have left?

17. Huan was given $2\frac{1}{4}$ h to complete his math test. However, he completed it in $1\frac{2}{3}$ h. How much time did he have to check his work? Give your answer in hours.

18. The height of a pole is 121 cm. $\frac{7}{11}$ of the pole is painted red and the rest is painted white. Find the length of the pole painted in white.

19. A banner was cut into two pieces. One piece was $2\frac{1}{5}$ m long and the other piece was $1\frac{3}{10}$ m longer than the first piece. How long was the banner at first?

20. Jerrell had \$108. He spent $\frac{1}{3}$ of his money on a ticket to a play and $\frac{1}{4}$ on a birthday gift. How much money does he have left?

BLANK

Name: _____   Date: _____

Test B   **40 min**

50

Score

Unit 3 Fractions

Section A (2 points each)
Circle the correct option: **A**, **B**, **C**, or **D**.

1.   What is the missing denominator?

$$\frac{5}{7} = \frac{65}{\boxed{}}$$

A 67   **B** 91   **C** 70   **D** 72

2.   Which of the following sets lists the equivalent fractions of $\frac{1}{7}$?

A $\frac{2}{14}, \frac{3}{21}, \frac{4}{28}$

B $\frac{1}{8}, \frac{1}{9}, \frac{1}{10}$

C $\frac{2}{8}, \frac{3}{9}, \frac{4}{10}$

D $\frac{2}{7}, \frac{4}{7}, \frac{6}{7}$

3. What is the missing number?

$$\frac{1}{5} + \frac{1}{5} + \frac{1}{5} + \frac{1}{5} + \frac{1}{5} + \frac{1}{5} = \boxed{?} \times \frac{2}{5}$$

   A 6   B 2

   C 3   D 4

4. What is the missing mixed number?

$$\boxed{?} + 1\frac{1}{3} = 3\frac{4}{9}$$

   A $2\frac{3}{9}$ B $\frac{20}{9}$

   C $2\frac{1}{9}$ D $2\frac{2}{3}$

5. In 2 years, Dalton's age will be $\frac{1}{4}$ his father's age. If Dalton's father is 46 years old now, how old is Dalton?

   A 10   B 12

   C 11   D 9

Section B (2 points each)

6. Write the missing number.

$$1\frac{3}{5} + 2\frac{1}{3} = 5 - \boxed{}$$

7. Write > , < or =.

$$\frac{19}{5} \underline{\hspace{2cm}} 4\frac{3}{5}$$

8. Arrange the following from the smallest to the greatest.

$$3\frac{1}{3} \qquad 4 \qquad 3\frac{1}{9} \qquad \frac{16}{3}$$

_____ , _____ , _____ , _____

smallest

9. Which three fractions below add up to 1?

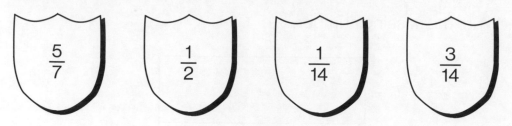

The three fractions are _____ , _____ , and _____ .

10. Fill in the blank with the correct number.

$$\frac{1}{8} + \frac{1}{8} + \frac{1}{8} + \frac{1}{8} + \frac{1}{8} + \frac{1}{8} = \underline{\hspace{2cm}} \times \frac{3}{8}$$

11. A picture measures 12 in. by $2\frac{1}{8}$ in. Express the area of this picture in its simplest form.

12. Express 30 months as a fraction of 2 years. Give your answer in its simplest form.

13. 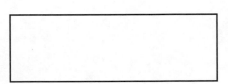 stands for a number. If $\frac{3}{4}$ of ⭐ is 27, what is $\frac{2}{9}$ of ⭐?

14. The mass of Box A is $\frac{3}{5}$ the mass of Box B. What is the total mass of the two boxes if the mass of Box B is 70 kg?

```



```

15. Mollie prepared $3\frac{2}{5}$ L of lemonade and 5 times as much orange juice as lemonade for a party. How many liters of orange juice did she prepare?

```



```

Section C (4 points each)

16. Jasmine made 8 jugs of hot cocoa for a winter festival. There were $1\frac{5}{6}$ liters of hot cocoa in each jug. How many liters of hot cocoa did she make altogether? Express your answer in its simplest form.

17. The total mass of 3 packages is 12 kg. Package A has a mass of $4\frac{1}{2}$ kg and Package B has a mass of $3\frac{3}{5}$ kg. What is the mass of Package C? Express your answer in its simplest form.

18.   Each kilogram of sand costs \$2 and each kilogram of cement costs

3 times as much as each kilogram of sand. Mr. Carrera bought $3\frac{1}{8}$ kg of

sand and $2\frac{1}{4}$ kg of cement. How much did he pay in all?

19.   Siti had 20 lb of clay. She used half of it to make some bowls and put
the rest equally into three containers. How much clay did she
put in each container? Express your answer in its simplest form.

20. Valde brought 5 bottles of drinking water on a camping trip. Each bottle contained $\frac{5}{8}$ L of water. At the end of his camping trip, he had $\frac{3}{4}$ L of water left in total. How much water did he drink during his camping trip?

BLANK

Name: _____   Date: _____

Test A **50 min**

60

Score

Continual Assessment 1

Section A (2 points each)
Circle the correct option: **A**, **B**, **C**, or **D**.

1.   In 2,739,465, the digit 3 is in the ___?___ place.

     A thousands   **B** ten thousands

     C hundred thousands   **D** millions

2.   What is the missing number?

   $7{,}000{,}000 + \boxed{?} + 700 + 70 = 7{,}070{,}770$

     A 7,000   **B** 70,000

     C 700,000   **D** 7,000,000

3.   What is the best way to round in order to find the estimate for
   $4{,}554 \times 34$?

     A $5{,}000 \times 40$ **B** $5{,}000 \times 30$

     C $4{,}000 \times 40$ **D** $4{,}000 \times 30$

4. Find the value of $34 - (6 + 9 \times 2) + 7$.

    A 81   **B** 17

    C 3   **D** 11

5. Which number is a common factor of 14 and 35?

    A 2   **B** 5

    C 7   **D** 14

6. What is the sum of the first 4 multiples of 6?

    A 18   **B** 24

    C 30   **D** 60

7. Which of the following is greater than $\frac{3}{8}$?

    A $\frac{3}{10}$ **B** $\frac{3}{12}$ **C** $\frac{3}{9}$ **D** $\frac{3}{5}$

8. Which one of the following does not give the answer of $1\frac{3}{4}$?

A  $7 \div 4$

B 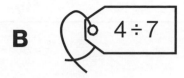 $4 \div 7$

C  $1\frac{1}{2} + \frac{1}{4}$

D 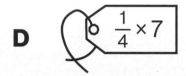 $\frac{1}{4} \times 7$

9. Nancy bought $\frac{4}{5}$ kg of flour. After she had used $\frac{2}{3}$ of the flour to make some pies, how much flour was left?

A $\frac{8}{15}$ kg

B $\frac{7}{15}$ kg

C $\frac{2}{15}$ kg

D $\frac{4}{15}$ kg

10. Given that $\frac{5}{6}$ of a number is 30, what is $\frac{1}{3}$ of the number?

A 36

B 24

C 18

D 12

Section B (2 points each)

11. Write 505 millions, 55 thousands, 5 tens and 5 ones in expanded form.

12. Complete the number pattern.

     624,307, _____, 864,307, 984,307, 1,104,307

13. What is the number that comes just before 1,000,000?

14. When an even number is rounded to the nearest thousand, it becomes 314,000. What is the largest possible even number that would be rounded to 314,000?

15.	Find the value of 10 to the fifth power.

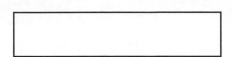

16.	The product of two numbers is 1,320. One number is 30. What is the other number?

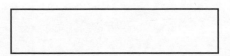

17.	Find the value of 30 + (36 − 12) ÷ 4 × 8.

18. What number does the letter A represent?

$$\frac{4}{5} = 2 - \frac{A}{5}$$

19. Divide 670 by 14.

20. $\frac{2}{5}$ of a number is 18. What is the number?

Section C (4 points each)

21. A rope was 5,280 cm long. It was cut into pieces of length 60 cm each. How many cuts were made?

22. A bookshop owner bought 600 comic books for $2,400. He then sold 200 comic books at $7 each and the remaining comic books at 2 for $11. How much money did he earn from the comic books?

23. 1,498 adults, 3,954 boys, and twice as many girls as boys visited Greenway Park in a week. How many people visited Greenway Park altogether?

24. The width of a rectangular garden is $9\frac{1}{4}$ m. Its length is $2\frac{3}{4}$ m longer than its width. What is the area of the garden?

25. $\frac{7}{10}$ of the chickens on a farm are hens. There are 30 roosters. How many chickens are there altogether?

Extra Credit

1.	X and Y each stand for a number. X is a 4-digit number with the digit 4 in its thousands place. The value of its hundreds digit is twice the value of its thousands digit. There are two zeros in X. Given that the sum of X and Y is 8,400, what number does Y stand for?

2. Karsen bought 2 similar shirts and 3 similar ties for $230. If he bought 3 similar shirts and 2 similar ties, he would need to pay $40 more. How much did each shirt cost?

BLANK

Test B

50 min

60

Score

Continual Assessment 1

Section A (2 points each)
Circle the correct option: **A**, **B**, **C**, or **D**.

1. In which of the following is the digit 6 in the ten thousands place?

    A 3,652,400   **B** 5,936,400

    C 6,870,450   **D** 9,065,380

2. The sum of two numbers is 682. If one number is 60 more than the other, what is the smaller number?

    A 622   **B** 311

    C 371   **D** 562

3. A number is a multiple of 7 and also a factor of 56. The number is __?__ .

    A 28   **B** 21

    C 8   **D** 1

4. At a train station, 90 passengers got off and 130 passengers boarded a train. If 340 passengers were on the train after that, how many passengers were there at first?

A 560

B 120

C 300

D 380

5. How many 50,000s are there in 10,000,000?

A two hundred

B two thousand

C twenty thousand

D two hundred thousand

6. The sum of two numbers is 30. One number is 6 more than the other number. Which number equation shows the value of the smaller number?

A $30 - 6 \div 2 = 27$

B $(30 - 6) \div 2 = 12$

C $30 - 6 \times 2 = 18$

D $(30 - 6) \div 2 + 6 = 27$

7.   Find the value of 12 + 8 ÷ 4 − 2 × 3.

   A 8   **B** 9

   C 12   **D** 36

8.   What is 48 ÷ 10 expressed as a fraction in its simplest form.

   A $4\frac{8}{10}$ **B** $4\frac{2}{5}$

   C $4\frac{4}{5}$ **D** $\frac{48}{10}$

9.   On the number line below, which letter represents $\frac{9}{4}$?

   A A   **B** B

   C C   **D** D

10. Vito had $\frac{1}{5}$ of his gift card balance left after spending $400. How much was his gift card worth originally?

    A $500   **B** $450

    C $320   **D** $120

Section B (2 points each)

11. In 900,682,700, which digit is in the ten thousands place?

12. Round each number to the nearest hundred and estimate the value of 7,968 × 243.

13. What is the largest even number that is smaller than 3,000,000 + 80,000 + 900?

14. Divide 803 by 37.

15. Show the prime factorization of 100 using exponents.

16. When a number is divided by 70, the quotient is 22 and the remainder is 50. What is the number?

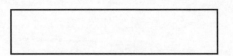

17. Find the value of $5 \times 6 + 12 \div 3 \times (4 + 5)$.

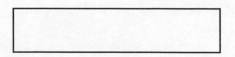

18. How much more is $\frac{9}{10}$ than $\frac{2}{5}$? Express your answer in its simplest form.

19. Add $\frac{2}{3}$ to the sum of $\frac{1}{6}$ and $\frac{7}{12}$. Express your answer in its simplest form.

20. $2\frac{3}{5}$ hours = _____ minutes

Section C (4 points each)

21. The chairs in a school auditorium were arranged in 15 rows with 28 chairs in each row. If these chairs were rearranged into rows of 20, how many rows of chairs would there be?

22. A math teacher wanted to purchase some calculators for her school district. If she bought 180 calculators, she would have $600 left. If she bought 156 calculators, she would have $1,080 left. What was the cost of each calculator?

23. Gemma spent $\frac{2}{7}$ of her money on a planner and $18 on a backpack. If she had $27 left, how much money did she have at first?

24. A rectangular living room is 8 m long and $4\frac{1}{6}$ m wide. The cost for carpeting the floor is $27 per m². What is the total cost for carpeting the entire floor of the room?

25. When $\frac{2}{7}$ of a wheelbarrow is filled with cement, it has a mass of 37 kg. When the same wheelbarrow is full of cement, it has a mass of 102 kg. What is the mass of the wheelbarrow when it is empty?

Extra Credit

1. The price of a museum ticket was $12 for an adult and $8 for a child. A group of people paid $900 altogether for their tickets. There were 30 more children than adults in the group. How many children were there?

2. A bakery had some tarts for sale. After 140 tarts were sold in the morning and $\frac{3}{8}$ of the remainder were sold in the afternoon, the bakery had $\frac{1}{2}$ of the tarts left. How many tarts did the bakery have at first?

Name: _____   Date: _____

Test A   **45 min**

50

Score

Unit 4 Multiply and Divide Fractions

Section A (2 points each)
Circle the correct option: **A**, **B**, **C**, or **D**.

1. Which of the following is the same as $\frac{1}{2} \div \frac{3}{10}$?

   A $\frac{3}{10} \times \frac{1}{2}$ **B** $\frac{1}{2} \times \frac{10}{3}$

   C $\frac{1}{2} \div \frac{10}{3}$ **D** $\frac{10}{3} \div \frac{1}{2}$

2. Pablo divided $\frac{3}{4}$ lb of beans equally into 4 portions. How many pounds of beans are in each portion?

   A $\frac{3}{8}$ lb   **B** 4 lb   **C** $1\frac{1}{4}$ lb   **D** $\frac{3}{16}$ lb

3. Makayla cuts $\frac{1}{3}$ of a pie into 3 pieces. What fraction of the whole pie is each piece?

   A $\frac{1}{6}$ **B** $\frac{1}{4}$ **C** $\frac{1}{9}$ **D** $\frac{1}{3}$

4. How many $\frac{5}{12}$s are there in $\frac{5}{6}$?

   A 2   **B** 5

   C 6   **D** 12

5. 4 friends shared $\frac{2}{3}$ of a watermelon. What fraction of the watermelon did each friend get?

   A $\frac{1}{6}$ **B** $\frac{1}{8}$ **C** $\frac{1}{12}$ **D** $\frac{3}{8}$

Section B (2 points each)

6.   Multiply. Write the answer in its simplest form.

$\dfrac{3}{4} \times \dfrac{5}{6} =$

7.   Multiply. Write the answer in its simplest form.

$\dfrac{3}{5} \times \dfrac{10}{11} =$

8.   Find the product. Write the answer in its simplest form.

$1\dfrac{3}{5} \times 10 =$

9.   Multiply $5\dfrac{1}{2}$ by $1\dfrac{3}{5}$. Write the answer in its simplest form.

Divide. Express each answer for questions 10 to 13 in its simplest form.

10. $3 \div \dfrac{1}{2} =$

11. $8 \div \dfrac{4}{5} =$

12. $\dfrac{3}{4} \div \dfrac{1}{8} =$

13. $\dfrac{3}{5} \div \dfrac{3}{10} =$

14. The perimeter of a square is $\frac{2}{5}$ m. What is the length of one side of the square? Express your answer in its simplest form.

15. The area of a mural is 10 ft² and its width is $\frac{3}{4}$ ft. What is the length of this mural? Express your answer in its simplest form.

Section C (4 points each)

16. Lea bought $\frac{4}{5}$ L of orange juice. She drank $\frac{3}{4}$ of it. How much

    orange juice did she drink? Give your answer in its simplest form.

17. Ryanne ate $\frac{1}{6}$ of a cake. She gave $\frac{3}{10}$ of the remainder to her sister.

    What fraction of the cake did she give to her sister? Give your

    answer in its simplest form.

18. 4 friends share $\frac{2}{3}$ of a pizza equally. What fraction of the pizza did each friend get? Give your answer in its simplest form.

19. $\frac{9}{10}$ kg of clay was shared equally among 3 children. What was the mass of clay each child received? Give your answer in its simplest form.

20. A jug contained $2\frac{1}{3}$ L of lemonade. Malik prepared $5\frac{1}{2}$ such jugs of lemonade. How many liters of lemonade did he prepare? Give your answer in its simplest form.

Name: _____   Date: _____

Test B

45 min

Unit 4 Multiply and Divide Fractions

Section A (2 points each)
Circle the correct option: **A**, **B**, **C**, or **D**.

1. How many $\frac{1}{12}$s are there in $\frac{1}{2}$?

    A 12   **B** 2

    C 6   **D** 24

2. The width of a rectangle is $\frac{3}{8}$ m. Its length is twice as long as its

    width. What is the area of the rectangle?

    A $1\frac{1}{8}$ m²   **B** $\frac{9}{32}$ m²   **C** $\frac{3}{4}$ m²   **D** $2\frac{1}{4}$ m²

3. Which of the following has the same value as $2\frac{3}{5} \times 4$?

A $6\frac{3}{5}$

B $2\frac{3}{5} \times 2\frac{3}{5} \times 2\frac{3}{5} \times 2\frac{3}{5}$

C $\frac{23}{5} \times \frac{1}{4}$

D $(2 \times 4) + (\frac{3}{5} \times 4)$

4. Macie made two identical pizzas and cut each pizza into a different number of equal pieces as shown below. She gave De'Andre one piece from each pizza. What fraction of the pizzas did De'Andre receive?

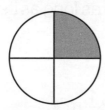 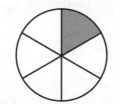

A $\frac{1}{5}$ B $\frac{1}{10}$ C $\frac{5}{12}$ D $\frac{5}{24}$

5. When Ángel pumped 30 L of gas into the fuel tank of his car, the fuel gauge moved from $\frac{1}{8}$ full to $\frac{3}{4}$ full. What was the capacity of the fuel tank?

A   48 L

B   40 L

C   60 L

D   64 L

Section B (2 points each)

Divide. Express each answer for problems 6 to 9 in its simplest form.

6. $\dfrac{3}{7} \div 12 = $

7. $9 \div \dfrac{3}{4} = $

8. $20 \div \dfrac{5}{9} = $

9. $\dfrac{2}{3} \div \dfrac{4}{9} = $

Divide. Express each answer for problems 10 and 11 in its simplest form.

10. $\dfrac{5}{6} \div \dfrac{5}{12} =$

11. The product of two numbers is $\dfrac{2}{3}$. One of the numbers is $\dfrac{4}{9}$. What is the other number? Express your answer in its simplest form.

12. How many $\frac{1}{10}$ kg portions of oatmeal can be made from $\frac{4}{5}$ kg?

13. The capacity of a teapot is $1\frac{4}{7}$ cups. The capacity of a kettle is $3\frac{1}{2}$ times that of the teapot. How many cups of water can the kettle hold?

14. The area of a regular piece of cloth is $\frac{2}{3}$ m². Its width is $\frac{4}{9}$ m.

What is its length? Give your answer in its simplest form.

15. $\frac{2}{3}$ of a sack of rice weighs 8 lb. How much does $\frac{1}{2}$ of a similar

sack of rice weigh?

Section C (4 points each)

16. Khalid cut a ribbon $1\frac{5}{9}$ yd long into 7 equal parts. What is the length of each part? Express your answer in its simplest form.

17. $\frac{4}{7}$ of the students in a school district are girls and $\frac{1}{3}$ of these girls wear glasses. Given that 264 girls wear glasses, how many students are there in the school district?

18. Harry read $\frac{1}{3}$ of a book on Saturday and $\frac{3}{4}$ of the remaining pages on Sunday. After that, he had 75 pages left. How many pages were there in the book?

19. $\frac{3}{5}$ of the balls in a box are red and $\frac{3}{8}$ of the remainder are black. If there are 147 black balls, how many balls are there in the box altogether?

20. There are some students rehearsing for a performance. $\frac{1}{3}$ of the students are practicing singing, $\frac{5}{6}$ of the remaining students are learning dance steps, and the rest of the students are playing musical instruments. What fraction of the students are playing musical instruments?

Name: _____ Date: _____

Test A **40 min**

Unit 5 Perimeter and Area

50

Section A (2 points each)
Circle the correct option: **A**, **B**, **C**, or **D**.

1. What is the area of the shaded triangle in the figure below?

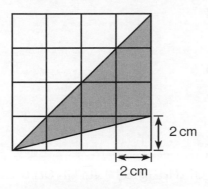

2 cm

2 cm

A $32\,cm^2$ **B** $24\,cm^2$

C $12\,cm^2$ **D** $6\,cm^2$

2. The figure below shows a triangle ABC. What is the base of triangle ABC given that its height is CE?

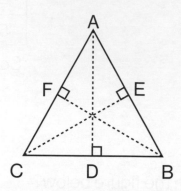

A BC

B AB

C AC

D AD

3. In the figure below, what is the height of triangle PQR given that the base is PQ?

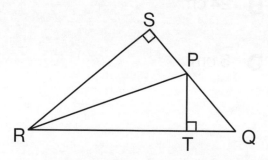

A PR

B PT

C RS

D QR

4.   In the figure below, what is the area of triangle MON?

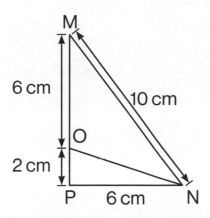

A 18 cm^2 **B** 24 cm^2

C 30 cm^2 **D** 36 cm^2

5.   What is the area of this parallelogram?

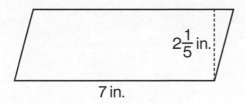

2$\frac{1}{5}$ in.

7 in.

A 14 in.^{2} **B** 15$\frac{2}{5}$ in.^{2}

C 7$\frac{1}{5}$ in.^{2} **D** 15$\frac{2}{7}$ in.^{2}

Section B (2 points each)

6. Find the area of the shaded triangle below.

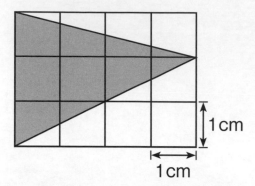

7. Find the area and perimeter of this figure. All the lines meet at right angles.

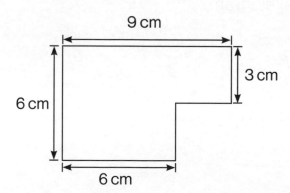

Perimeter = _____

Area = _____

8. Find the area and perimeter of this figure. All the lines meet at right angles.

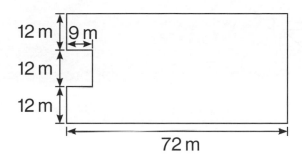

Perimeter = _____

Area = _____

9. Which of the following triangles does not have its height drawn correctly?

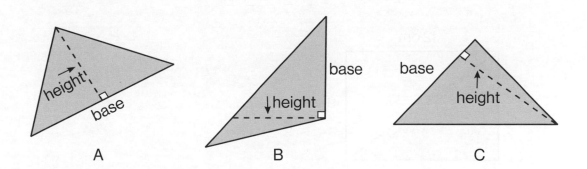

A

B

C

10. In the figure below, what is the area of the triangle ABC? Give your answer in its simplest form.

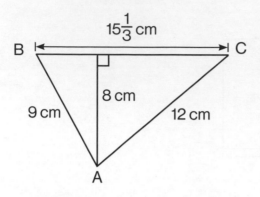

11. The figure below shows a rectangle measuring 12 cm by 9 cm. What is the area of the shaded part in this rectangle?

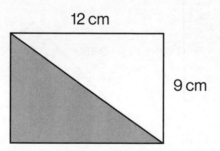

12 cm

9 cm

12. The length of a rectangle below is three times its width. Its perimeter is 112 m.

Perimeter = 112 m

The length of the rectangle is _____.

13. Find the area of the shaded part of this rectangle.

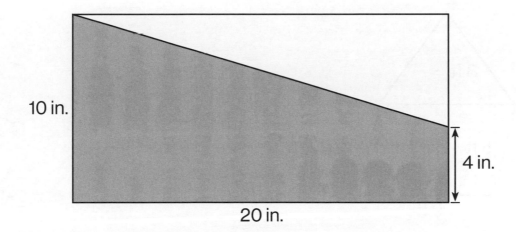

10 in.

4 in.

20 in.

The area of the shaded part is _____.

14. The figure below is made up of two squares. One has sides that are 20 meters long and the other has sides that are 12 meters long.

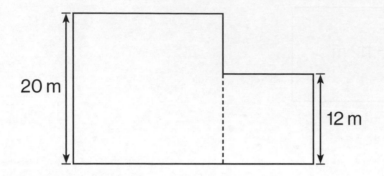

The perimeter of this figure is _____.

15. Find the area of the figure below.

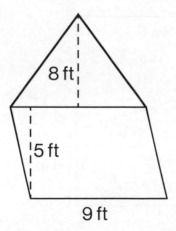

8 ft

5 ft

9 ft

Section C (4 points each)

16. The figure below shows a rectangle measuring 28 cm by 23 cm. What is the area of the unshaded parts?

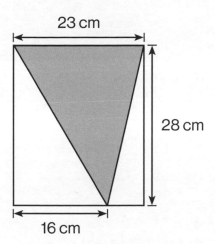

17. The figure below shows a rectangle measuring 30 cm by 16 cm. What is the area of the shaded part?

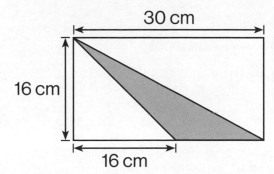

18. The figure below is made up of 2 squares with sides 27 cm and 21 cm respectively. What is the area of the shaded part?

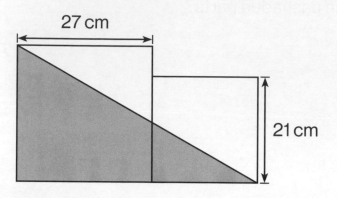

27 cm

21 cm

19. A rectangular garden measures 24 m by 10 m. A 6-meter square area is covered with some plants. What is the area of the garden that is not covered by the plants?

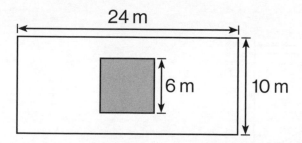

24 m

6 m

10 m

20. The perimeter of a rectangular swimming pool is 52 m. Its width is 8 m shorter than its length. What is the area of the swimming pool?

Perimeter = 52 m

BLANK

Name: _____ Date: _____

Test B   40 min

50
Score

Unit 5 Perimeter and Area

Section A (2 points each)
Circle the correct option: **A**, **B**, **C**, or **D**.

1.   What is the area of the shaded triangle in the figure below?

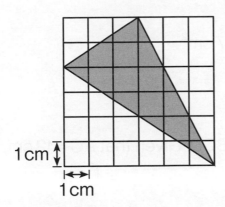

1 cm
1 cm

A 24 cm^2 **B** 18 cm^2

C 15 cm^2 **D** 12 cm^2

2. The figure below is made up of 2 rectangles. What is the area of the unshaded parts?

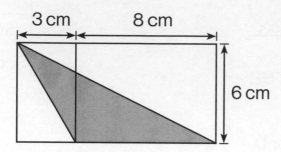

3 cm   8 cm

6 cm

A 24 cm^2

B 33 cm^2

C 42 cm^2

D 57 cm^2

3. The figure below shows a square ABCD. Given that AC = 12 cm, what is the area of the square?

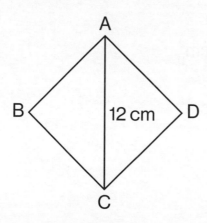

A

B   12 cm   D

C

A 36 cm^2

B 72 cm^2

C 108 cm^2

D 144 cm^2

Tests 5A

4. The figure below is made up of 2 rectangles of the same width. What is the area of the shaded part?

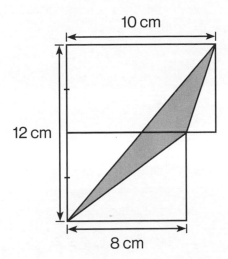

10 cm

12 cm

8 cm

A 48 cm²

B 42 cm²

C 24 cm²

D 18 cm²

5. The figure below shows a rectangle ABCD measuring 10 cm by 6 cm. What is the total area of the two shaded triangles?

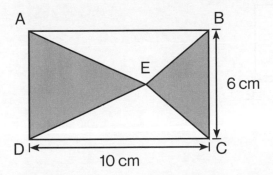

A B

E

6 cm

D C

10 cm

A 30 cm²

B 60 cm²

C 32 cm²

D 16 cm²

Section B (2 points each)

6. Find the area of the shaded part below.

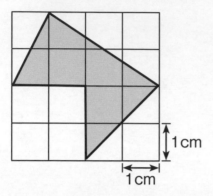

1 cm

1 cm

Area of shaded part = _____

7. The figure below is made up of two identical rectangles ABCD and BCEF. The area of triangle ACE is $\frac{1}{2}$ the area of which rectangle?

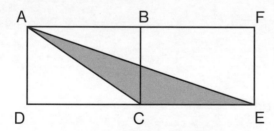

8.   The figure below consists of squares with 4 cm sides. Find the area of the shaded parts.

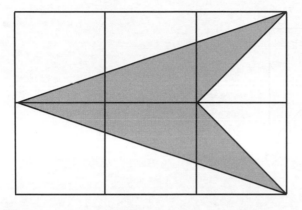

9.   The figure below is made up of two parallelograms. Find its area.

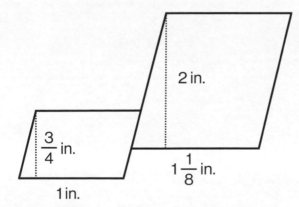

10. The figure below shows a rectangle with an area of 108 cm². Given that DE = EF = FC, find the total area of the shaded parts in the figure.

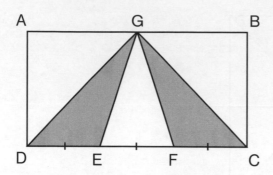

[ ]

11. Find the area and perimeter of the following figure.

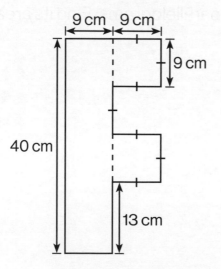

Perimeter = _____

Area = _____

12. In the figure below, a rectangular piece of paper measuring 26 cm by 18 cm is folded at corner D in such a way that BD is $\frac{1}{3}$ of its width. Find the area of triangle ABD.

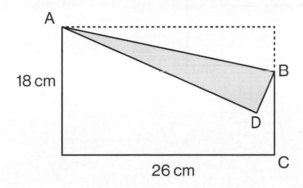

18 cm

26 cm

13. The figure to the right is made up of 6 identical triangles. Given that the perimeter of square EFGH is 72 cm, find the area of the whole figure.

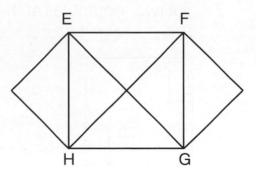

14. Two rectangular pieces of cardboard are pieced together to form a square of perimeter 112 cm as shown below.

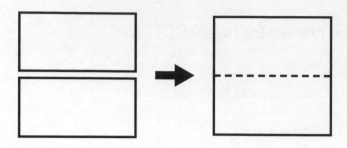

The area of each rectangular piece of cardboard is

_____.

15. Serena has a square piece of paper with an area of 81 cm². She cuts out two rectangles at the two corners as shown below.

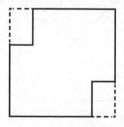

The perimeter of the remaining piece of paper is _____.

Section C (4 points each)

16. Tyrone cut the shaded part out from a triangular piece of cardboard as shown below. What is the area of the cardboard that is left? Give your answer in its simplest form.

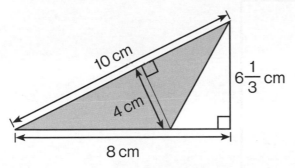

17. Richard ran around a rectangular field 4 times. He ran a total of 1,000 m. The width of the field was 35 m. What was the length of the field?

18. A painting measuring 9 ft by 7 ft hangs in the center of a square wall. The square wall has an area of 900 ft^2. How far above the ground is the bottom edge of the painting?

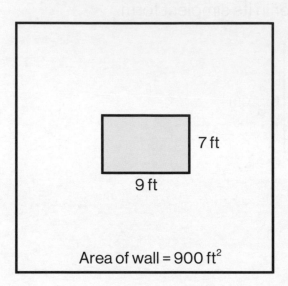

7 ft

9 ft

Area of wall = 900 ft^2

Tests 5A

4 identical right triangles are used to form a square as shown below. Use this illustration to answer questions 19 and 20.

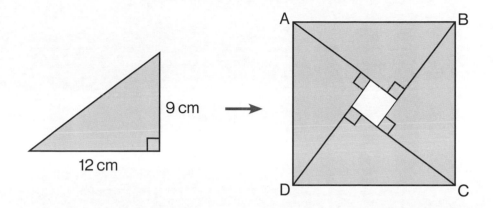

19. What is the area of square ABCD?

20. What is the length of square ABCD?

BLANK

Name: _____ Date: _____

Test A  **30 min**

Unit 6 Ratio

Section A (2 points each)
Circle the correct option: **A**, **B**, **C**, or **D**.

1. The figure below is made up of identical squares. What is the ratio of the number of shaded squares to the number of unshaded squares?

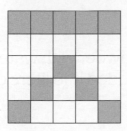

    A 2:3   **B** 2:5

    C 3:2   **D** 5:2

2. Martín is 8 years old. He is 2 years older than his sister. What is the ratio of Martín's age to his sister's age?

    A 4:3   **B** 4:14

    C 2:7   **D** 1:4

3. Which one of the following is **<u>not</u>** an equivalent ratio of 12 : 42?

12 : 42

A 6 : 21

B 4 : 14

C 2 : 7

D 1 : 4

4. Peggy is preparing some paint by mixing dry powder with water in the ratio of 2 : 3. How much dry powder is needed if 18 liters of water is used?

A 6 liters

B 12 liters

C 27 liters

D 90 liters

5. A box contains red beans and black beans in the ratio 3 : 5. If there are 10 kg more black beans than red beans, what is the total mass of beans in the box?

A 15 kg

B 25 kg

C 40 kg

D 80 kg

Section B (2 points each)

6. Write the correct ratio for the following.

The ratio of the number of model cars to the number of model ships is _____ : _____.

7. Write the correct ratio for the following.

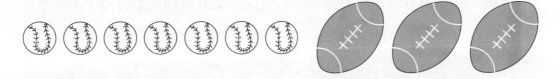

a) The ratio of the number of baseballs to the number of footballs is _____ : _____.

b) The ratio of the number of footballs to the number of baseballs is _____ : _____.

c) The ratio of the total number of baseballs and footballs to the number of baseballs is _____ : _____.

8. Express each of the following ratios in the simplest form.

a) 8 : 4 = _____ : _____

b) 3 : 9 = _____ : _____

9. Write the correct ratio for the following.

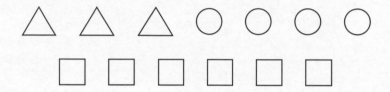

The ratio of the number of triangles to the number of circles to the number of squares is _____ : _____ : _____.

10. Express the following ratio in its simplest form.

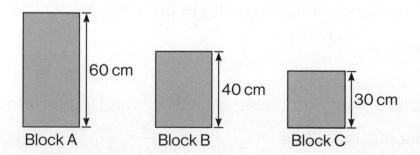

Block A   Block B   Block C

The ratio of Block A's height to Block B's height to Block C's height is _____ : _____ : _____.

11. Complete the following equivalent ratios.

    a) $3:4:1 = \underline{\hspace{1cm}}:8:\underline{\hspace{1cm}}$

    b) $5:4:3 = \underline{\hspace{1cm}}:\underline{\hspace{1cm}}:12$

12. Express each of the following ratios in the simplest form.

    a) $16:20:8 = \underline{\hspace{1cm}}:\underline{\hspace{1cm}}:\underline{\hspace{1cm}}$

    b) $24:12:40 = \underline{\hspace{1cm}}:\underline{\hspace{1cm}}:\underline{\hspace{1cm}}$

13. How many more stars must be shaded so that the ratio of the number of shaded stars to the total number of stars is $2:3$?

14. Tanner is 12 years old. His brother is 6 years older. What is the ratio of Tanner's age to his brother's age? Express your answer in its simplest form.

_____ : _____

15. Tree A is 5 times as tall as Tree B. What is the ratio of the height of Tree A to the height of Tree B? Express your answer in its simplest form.

_____ : _____

Section C (4 points each)

16. Hanh spent $25 and had $15 left. What was the ratio of the amount of money she spent to the total amount of money she had at first? Express your answer in its simplest form.

17. There are 40 students signed up for an art camp. 24 of them attend camp during Week 1 and the rest attend camp during Week 2.

   a) What is the ratio of the number of campers in Week 1 to the number of campers in Week 2? Express your answer in its simplest form.

   b) What is the ratio of the number of campers in Week 2 to the total number of students signed up for the art camp? Express your answer in its simplest form.

18. The table below shows the amount of money 4 friends saved in a month. Write the correct ratios in the blanks. Express the ratios in their simplest form.

Names	Arika	Ben	Cari	Della
Amount Saved	$16	$9	$8	$12

a) The ratio of Arika's savings to Ben's savings is _____ : _____.

b) The ratio of Della's savings to Cari's savings is _____ : _____.

(c) The ratio of Ben's savings to Della's savings is _____ : _____.

(d) The ratio of Cari's savings to the total savings of the 4 friends is _____ : _____.

19. Devin spent 15 minutes doing his math homework and 2 hours doing his English homework. What was the ratio of the time he spent on his math homework to the time he spent on his English homework? Express the ratio in the simplest form.

20. Three children shared a sum of money in the ratio 6 : 4 : 5. Given that the biggest share was $30, find the sum of money.

BLANK

Name: _____   Date: _____

Test B **30 min**

50

Unit 6 Ratio

Section A (2 points each)
Circle the correct option: **A**, **B**, **C**, or **D**.

1.   In the figure below, how many more triangles must be shaded to
      make the ratio of the number of shaded triangles to the total
      number of triangles 3 : 5?

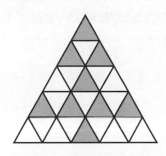

     A 15   **B** 10

     C 9   **D** 6

2.   Which one of the following is an equivalent ratio of 2 : 7?

     A 2 : 14   **B** 4 : 14

     C 4 : 9   **D** 9 : 4

3. A box contains some red, yellow, and blue beads. The ratio of the number of red beads to the number of yellow beads is 2 : 3. The ratio of the number of yellow beads to the number of blue beads is also 2 : 3. What is the ratio of the number of red beads to the number of blue beads in the box?

A 2 : 3

B 3 : 2

C 4 : 9

D 9 : 4

4. Andrea is 12 years old and her cousin is twice as old as she is. What will be the ratio of Andrea's age to her cousin's age in 8 years' time?

A 2 : 3

B 5 : 8

C 2 : 1

D 8 : 5

5. What is the ratio of the length of A to the length of B to the length of C?

A 8 : 4 : 18

B 4 : 8 : 18

C 4 : 6 : 8

D 8 : 4 : 6

Section B (2 points each)

6. Express the ratios in their simplest form.

a) The ratio of the number of apples to the number of bananas

is _____ : _____.

b) The ratio of the number of bananas to the number of apples

is _____ : _____.

c) The ratio of the number of bananas to the total number of

fruits is _____ : _____.

7. Express the ratio in its simplest form.

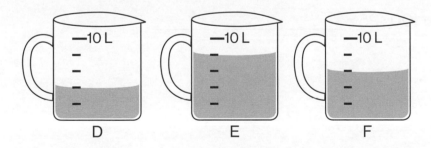

The ratio of the volume of water in container D to that in

container E to that in container F is _____ : _____ : _____.

8. Complete the equivalent ratios.

$11 : 5 : 3 = \underline{\hspace{2cm}} : 25 : \underline{\hspace{2cm}}$

$10 : 6 : 9 = \underline{\hspace{2cm}} : \underline{\hspace{2cm}} : 54$

9. Express the following ratios in their simplest form.

a) $30 : 6 : 18 = \underline{\hspace{2cm}} : \underline{\hspace{2cm}} : \underline{\hspace{2cm}}$

b) $21 : 14 : 63 = \underline{\hspace{2cm}} : \underline{\hspace{2cm}} : \underline{\hspace{2cm}}$

c) $72 : 48 : 60 = \underline{\hspace{2cm}} : \underline{\hspace{2cm}} : \underline{\hspace{2cm}}$

d) $75 : 100 : 25 = \underline{\hspace{2cm}} : \underline{\hspace{2cm}} : \underline{\hspace{2cm}}$

10. The table below shows the masses of 5 boxes. Write the correct ratios in the blanks.

Box	A	B	C	D	E
Mass	4 kg	7 kg	6 kg	3 kg	5 kg

a) The ratio of the mass of Box A to the mass of Box D is

_____ : _____.

b) The ratio of the mass of Box C to the mass of Box E is

_____ : _____.

(c) The ratio of the mass of Box B to the total mass of the

5 boxes is _____ : _____.

(d) The ratio of the total mass of the 5 boxes to the mass of

Box A is _____ : _____.

11. What is the ratio of the perimeter of triangle A to the perimeter of triangle B to the perimeter of triangle C? Express the ratio in its simplest form.

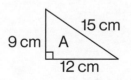

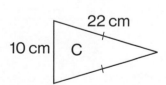

_____ : _____ : _____

12. In a box, the number of red pens is 3 times the number of blue pens. The number of black pens is twice the number of blue pens. What is the ratio of the number of red pens to the number of blue pens to the number of black pens?

_____ : _____ : _____

13. What is the ratio of the area of Rectangle A to the area of Square B? Express the ratio in its simplest form.

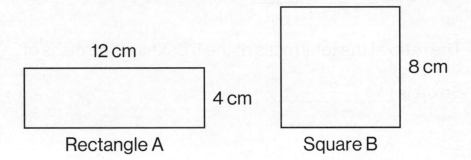

12 cm

4 cm

Rectangle A

8 cm

Square B

_____ : _____

14. The ratio of the base of a triangle to its height is 2:3. The height of the triangle is 12 ft. Find its area.

15. A rope is cut into three pieces in the ratio 5 : 2 : 3. The length of the shortest piece is 84 cm. What is the length of the longest piece?

Section C (4 points each)

16. The ratio of the number of mangoes to the number of avocados to the number of papayas is 3 : 5 : 9. There are 30 avocados. How many papayas are there?

17. Monette mixes 2 liters of fruit punch with every 5 liters of sparkling water to make a drink. How many liters of fruit punch are needed if she uses 30 liters of water?

18. At the animal shelter, the ratio of the number of cats to the number of dogs is 7 : 3. There are 36 dogs at the shelter.

   a) How many cats are there at the animal shelter?

   b) How many cats and dogs are there at the animal shelter?

19. The ratio of three sides of a triangle is 4 : 5 : 3. If the perimeter of the triangle is 60 cm, how long is the longest side of the triangle?

20. A florist had some roses and lilies in the ratio 2 : 5. If she got 30 more roses, the ratio of the number of roses to the number of lilies would be 8 : 15. How many roses and lilies did the florist have altogether before getting any more flowers?

Name: _____   Date: _____

Test A

70 min

80

Score

Continual Assessment 2

Section A (2 points each)
Circle the correct option: **A**, **B**, **C**, or **D**.

1.   What is the missing number?

   $700{,}506 = \boxed{\quad ? \quad} + 500 + 6$

   A 700   **B** 7,000

   C 70,000   **D** 700,000

2.   In 8,439,765, the digit 4 is in the _____ place.

   A thousands   **B** ten thousands

   C hundred thousands   **D** millions

3.   Which of the following is a composite number?

   A 7   **B** 11

   C 22   **D** 23

4.   5 lamps and a fan cost $330 altogether. If the fan costs $80, what is the cost of each lamp?

   A $25   **B** $50

   C $175   **D** $250

5.   What is the prime factorization of 108?

   A $3 \times 3 \times 3 \times 4$ **B** $2 \times 2 \times 2 \times 3 \times 3$

   C $2 \times 2 \times 3 \times 3$ **D** $3 \times 3 \times 4 \times 4$

6.   Calculate $15 + 30 \div (15 - 9) \times 4$.

   A 80   **B** 35

   C 69   **D** 21

7. Which one of the following is an equivalent fraction of $\frac{3}{7}$?

  A $\frac{3}{8}$ **B** $\frac{6}{7}$ **C** $\frac{15}{35}$ **D** $\frac{6}{12}$

8. What is the value of $2\frac{5}{12} + 3\frac{1}{4}$ expressed as a mixed number in its simplest form?

  A $5\frac{6}{12}$ **B** $5\frac{9}{12}$ **C** $5\frac{2}{3}$ **D** $5\frac{8}{12}$

9. In $\frac{3}{21} = \frac{4}{\triangle}$, what does $\triangle$ stand for?

  A 28   **B** 12

  C 7   **D** 84

10. Winnie had $\frac{1}{5}$ of her savings left after spending $200. What was her savings to start with?

    A $800   **B** $250

    C $1,000   **D** $50

11. The figure below is made up of 4 identical squares. Express the shaded triangle as a fraction of the figure.

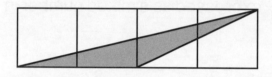

    A $\frac{1}{4}$ **B** $\frac{3}{8}$ **C** $\frac{5}{12}$ **D** $\frac{3}{4}$

12. What is the area of the shaded part?

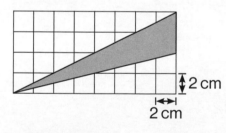

2 cm

2 cm

    A 32 cm^2 **B** 24 cm^2

    C 16 cm^2 **D** 8 cm^2

13. If AB is the base of the shaded triangle below, which of the following is its height?

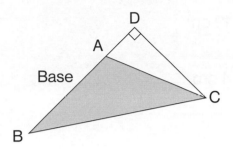

A AC

B BC

C DA

D DC

14. What is the area of the parallelogram?

78 in.

29 in.

A 2,192 in.^{2}

B 2,162 in.^{2}

C 858 in.^{2}

D 2,262 in.^{2}

15. The ratio of the sides of a triangle is $4 : 3 : 2$. The perimeter of the triangle is 36 cm. What is the length of the longest side?

    A 16 cm   **B** 12 cm

    C 8 cm   **D** 4 cm

Section B (2 points each)

16. Write three million, thirty-five thousand, two hundred fifty in numerals.

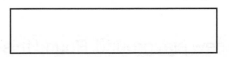

17. Write the missing number.

88,345,087 = 80,000,000 + 8,000,000 + 40,000 +

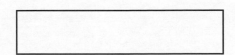 + 80 + 7

18. Find the value of 6 to the third power.

19. Write the missing number.

$$60 \times 1{,}500 = 15 \times \boxed{} \times 2$$

20. A sum of $67,436 was raised for a new playground. Round this amount of money to the nearest $1,000.

21. Round each number to the nearest hundred and estimate the value of 4,957 × 336.

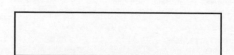

22. Cross out (✕) the numbers that are **not** multiples of 6.

| 6 | 10 | 12 | 16 | 20 | 24 |

23. Write the missing number.

$35 - 5 \times (42 \div 7) =$ _____ $\div 100$

24. Fill in the blank with the correct number.

$\dfrac{2}{7} + \dfrac{2}{7} + \dfrac{2}{7} + \dfrac{2}{7} =$ _____ $\times \dfrac{4}{7}$

25. The area of a rectangular hall is 1,260 m². Its width is 37 m. What is the perimeter of the hall? Give your answer in its simplest form.

26. Name the base related to the given height in each triangle.

a)

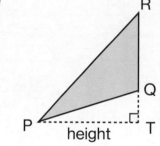

b)

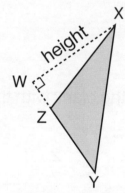

27. The figure below is formed by two squares and a right triangle. The areas of the two squares are 144 cm² and 169 cm² respectively. What is the area of the triangle?

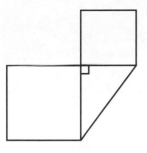

28. In the rectangle below, what is the area of the shaded triangle? Give your answer in its simplest form.

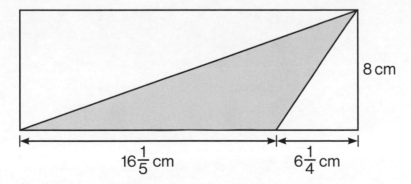

29. Write the missing number.

$$\boxed{} : 12 = 36 : 54$$

30. The ratio of the length to the width of a rectangle is 3 : 1. The length of the rectangle is 12 cm. What is its perimeter?

$$\boxed{}$$

Section C (4 points each)

31. There are 83 horses, 792 goats and 546 sheep on a ranch. How many animals are there on the ranch? Round your answer to the nearest 100.

32. Angela baked 554 cupcakes and packed them into boxes of 20 cupcakes each. If 14 cupcakes were left unpacked, how many boxes of cupcakes did she pack?

33. A board is cut into two pieces. One piece is $2\frac{1}{4}$ m long and the other piece is $\frac{3}{5}$ m longer. How long was the board at first? Give your answer in its simplest form.

34. Ryo bought $\frac{4}{5}$ kg of cashew nuts. He ate $\frac{3}{8}$ of them. What was the mass of cashew nuts he ate? Give your answer in its simplest form.

35. In a study tracking 40 wolves in a forest, $\frac{2}{5}$ of the wolves were male. After 2 more male wolves joined the pack and 6 females left the pack, what fraction of the wolves were male? Express your answer in its simplest form.

Extra Credit

1. Jimena and Liam each had some money. After Jimena spent $\frac{2}{5}$ of her money and Liam spent $\frac{6}{7}$ of his money, Jimena had 3 times as much money as Liam. Liam spent $150. How much did Jimena spend?

2. Noelani and Sun-mi folded some paper cranes in the ratio 3 : 7. After Noelani folded 35 more paper cranes, the ratio became 2 : 3. How many paper cranes did Noelani fold altogether?

BLANK

Name: _____   Date: _____

Test B   **70 min**

80

Score

Continual Assessment 2

Section A (2 points each)
Circle the correct option: **A**, **B**, **C**, or **D**.

1. What is the missing number?

    $18 \times \boxed{\,?\,} + 500 + 70 = 180,570$

    A 1,000

    B 10,000

    C 100,000

    D 10

2. What is the value of $24 \times 100,000 + 8 \times 1,000 + 5 \times 100 + 3 \times 10$?

    A 24, 853

    B 240,853

    C 248,530

    D 2,408,530

3. The price of a computer is $2,188. Round the price to the nearest thousand dollars and estimate the cost of 9 such computers.

    A $19,800

    B $19,692

    C $18,000

    D $20,000

4. A number gives the same answer when rounded to the nearest ten, hundred, or thousand. Which of the following can be that number?

   A 2,988　　　　　　　　B 2,990

   C 2,998　　　　　　　　D 2,899

5. What is the sum of the smallest prime number and the smallest composite number?

   A 3　　　　　　　　　　B 11

   C 6　　　　　　　　　　D 8

6. Abdullah has 128 toy cars. He packs all of them into boxes which hold up to 13 toy cars each. Which of the following cannot be the total number of boxes he used?

   A 8　　　　　　　　　　B 10

   C 15　　　　　　　　　　D 20

7. Adelina earns $400 in 8 days. How many days will it take her to earn $6,000?

   A 120   **B** 50

   C 12   **D** 5

8. What is the value of $25 + 60 \div (30 - 25) \times 2$?

   A 49   **B** 37

   C 34   **D** 4

9. At a farmer's market, 147 balloons were distributed among the children that visited that day. Each child received 4 balloons and 3 balloons were left over. How many children visited the farmer's market that day?

   A 36   **B** 37

   C 39   **D** 40

10. Which of the following has the same value as $\frac{3}{8} \div 12$?

   A $\frac{3}{8} \times 4$ **B** $\frac{1}{8} \times 4$ **C** $\frac{3}{8} \div 4$ **D** $\frac{1}{8} \div 4$

11. What is the fraction exactly halfway between $\frac{1}{5}$ and $\frac{4}{5}$?

   A $\frac{1}{2}$ **B** $\frac{3}{5}$ **C** $\frac{2}{5}$ **D** $\frac{1}{4}$

12. After Mikoda spent $60 on a pair of shoes and $\frac{3}{5}$ of the remainder of his money on a T-shirt, he had $44 left. How much money did he have at first?

   A $210   **B** $170

   C $110   **D** $50

13.　What is the area of the triangle below?

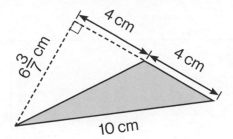

A $12\frac{6}{7}$ cm^2 **B** $25\frac{5}{7}$ cm^2

C $32\frac{1}{7}$ cm^2 **D** 40 cm^2

14.　Kini and Evelyn shared $96. Kini received $16 more than Evelyn. What was the ratio of Evelyn's share to Kini's share?

A 5 : 2   **B** 7 : 5

C 5 : 12   **D** 5 : 7

15.　A basket containing a dozen eggs is dropped from a height. Which one of the following **cannot** be the ratio of the number of broken eggs to those not broken?

A 4 : 3   **B** 2 : 1

C 1 : 5   **D** 5 : 7

Section B (2 points each)

16. Write 180,350,123 in expanded form.

17. 999,900 is [   ] less than 1 million.

18. In 8,790,436,

    the digit 8 is in the _____ place,

    the value of the digit 9 is _____, and

    the digit _____ is in the thousands place.

19. What is the smallest number that can be divided evenly by 3, 6 and 12? Circle it.

3   12   9   216

20.   What is the greatest common factor of 30 and 50?

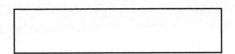

21.   Fill in the blank with >, < or =.

5^3 ____ 3^5

22.   Show the prime factorization of 800 using exponents.

23. When a number is divided by 73, the quotient is 46 and the remainder is 20. What is the number?

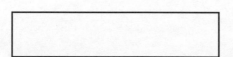

24. Find the value of $33 - (3 + 4) \times 3 - 10 \div 5$.

25. A bucket was $\frac{2}{3}$ full. When 760 ml of water was poured out, it was $\frac{1}{4}$ full of water. What was the capacity of the bucket?

26. All the lines in the following figure meet at right angles. Find the unknown marked length.

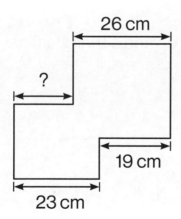

26 cm

?

19 cm

23 cm

27. A piece of wire 3 m long was used to form 4 equilateral triangles as shown below. The length of PQ is 37 cm. Find the length of the wire left.

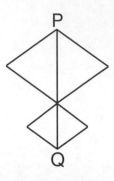

28. The figure below is made up of two squares and a triangle. The areas of the squares are 144 cm² and 324 cm² respectively. What is the area of the triangle?

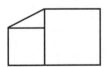

29. A rectangular piece of paper, colored on one side, is folded to form the shape shown below. The perimeter of the piece of paper is 44 cm.

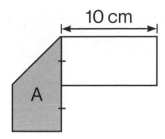

10 cm

A

a) What is the width of the piece of paper?

b) What fraction of the whole piece of paper is A? Give your answer in its simplest form.

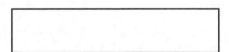

30. A 20-meter piece of string is cut into two pieces. The shorter piece is 8 meters long.

   a)   What is the ratio of the length of the longer piece to the length of the shorter piece? Give your answer in its simplest form.

   _____ : _____

   b)   What is the ratio of the length of the original string before it was cut to the length of the longer piece? Give your answer in its simplest form.

   _____ : _____

Section C (4 points each)

31. Three classes collected a total of 3,600 books. Class A collected 390 books more than Class B. Class B collected twice as many books as Class C. How many books did Class A collect?

32. Saul is 15 years old. Kade is 38 years older than Saul. In how many years' time will Kade be three times as old as Saul?

33. Lata spent $\frac{3}{5}$ of her vacation budget on food and $\frac{1}{6}$ of the remainder on transportation. If she had $380 left, how much was her vacation budget?

34. Robert cut a triangular piece out from a square tile with sides 18 cm long as shown below. What was the area of the tile left?

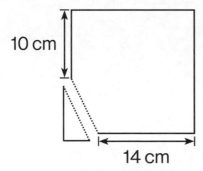

10 cm

14 cm

35. The ratio of the number of red beads to the number of blue beads in a jar was 3 : 2. Some red beads were taken out, and the new ratio of the number of red beads to the number of blue beads became 2 : 5. If there were 300 red beads at first, how many red beads were taken out?

Extra Credit

1. In the figure below, the area of rectangle ABCD is 120 cm^2.
    Given that AE = EF = FG = GH = HD, find the total area of the
    shaded parts.

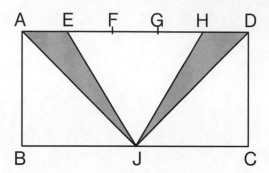

        Tests 5A

2. Richard has some red marbles, green marbles, and blue marbles in the ratio 7 : 3 : 12. If Richard replaces 24 blue marbles with the same number of red marbles, the number of red marbles will be 3 times the number of green marbles. How many green marbles does Richard have?

BLANK

Answer Key and Detailed Solutions

Unit 1　Test A

1. A
2. D
3. B
4. D
5. C

2	48
2	24
2	12
2	6
3	3
	1

6. five million, forty-three thousand, seven hundred eight

7. 900,000,000 + 80,000,000 + 2,000,000 + 600,000 + 50,000 + 3,000 + 7

8. 210,000,000

9. 2,543,090

10. 9,000
    ten thousands
    1,000,000
    4
    21,000

11. 30,295 and 30,349

12. 2,017,380,　2,107,308,　2,701,038

13a. 4, 5

13b. 8, 10

14. Multiples of 6: 6, 12, 18 …
    Multiples of 9: 9, 18 …
    Answer: Lowest Common Multiple is 18

15. Factors of 20: 1, 2, 4, 5, 10, 20
    Factors of 45: 1, 3, 5, 9, 15, 45
    Answer: Greatest Common Factor is 5

16. $1,823,450 ≈ $1,823,000
    Answer: $1,823,000

17. The common multiples of 3 and 7 which is less than 50 are 21 and 42.
    Answer: 42 m

18. 4,820 ≈ 5,000
    5,000 ÷ 20 = 250
    Answer: 250 cans

19. 888 × 500 = 444,000
    Answer: 444,000 sheets of paper

20. $70,000 ÷ 5,000 = $14
    Answer: $14

Unit 1   Test B

1. B

2. B

3. C

4. A
   $36 = 2 \times 2 \times 3 \times 3 = 2^2 \times 3^2$

5. B
   Factors of 8: 1, 2, 4, 8
   Factors of 12: 1, 2, 3, 4, 6, 12
   Multiples of 8: 8, 16, 24 ...
   Multiples of 12: 12, 24 ...
   GCF = 4, LCM = 24

6. $528,000 \div 1,000 = 528$
   Answer: 528

7a. 10^3, 10^3

7b. 2×10^3, 2×10^3

8. Multiples of 3: 3, 6, 9, 12, 15, 18, 21 ...
   Multiples of 7: 7, 14, 21 ...
   Answer: 21

9. 4
   5,000,000
   3,000
   hundred thousands
   4

10a. 865,421

10b. 865,412

10c. 124,568

10d. 124,685

11. 8,204,315,   8,025,134,   8,024,513,
    8,024,135

12. $2^2 \times 3^2$

13. $3 \times 6 = 18$
    $18 \div 2 = 9$
    Answer: 9

14. $57 - 1 = 56$
    Answer: 56

15. 6,000 more than 4,000 is 10,000.
    6,000 more than 12,000 is 18,000.
    6,000 more than 44,000 is 50,000.
    6,000 more than 994,000 is 100,000.
    Answer: 994,000

16. $4,950,600 \approx 5,000,000$
    $3,138,500 \approx 3,100,000$
    $5,000,000 + 3,100,000 = 8,100,000$
    Answer: 8,100,000

17. The common multiple of 6 and 9
    which is less than 30 is 18.
    Answer: $18

18. $280 \approx 300$
    $170 \approx 200$
    $300 \times 200 = 60,000$
    Answer: 60,000 ft^2

19. $614 \times \$3,000 = \$1,842,000$
    Answer: $1,842,000

20. $\$3,450 + \$1,548 - \$3,920$
    $\approx \$3,000 + \$2,000 - \$4,000$
    $= \$1,000$
    Answer: $1,000

Unit 2   Test A

1.   D

      $18 - 12 \div (3 \times 2) + 2$

      $= 18 - 12 \div 6 + 2$

      $= 18 - 2 + 2$

      $= 18$

2.   A

3.   C

4.   A

5.   D

6. $30 + 24 \div 4 - 2$

      $= 30 + 6 - 2 = 34$

7. $36 \div 9 + 3 \times (8 - 6)$

      $= 4 + 3 \times 2$

      $= 4 + 6$

      $= 10$

8. $5 \times \{ [8 \div 4] + 7 \} = 5 \times \{ 2 + 7 \} =$

      $5 \times 9 = 45$

      Answer: 45

9. $6 \div (3 \times 2) + 2 = 3$

10. $(\underline{2} \times 16) + (9 \times \underline{16})$

11. $(100 \times \underline{5}) - (\underline{1} \times 5)$

12.   15,925

13.   312

14. $1,033 - 65 = 968$

      Answer: 968 lb

15. $4,680 \div 60 = 78$

      Answer: 78 m

16. $(\$750 - \$390) \div 2 = \$180$

17. $68 \div 2 = 34$

      $34 \times 12 = \$408$

      Answer: He received $408

18. $336 \div 8 = 42$

      $42 - 35 = 7$

      Answer: 7 boxes were left unsold

19. $8,454 \div 6 = 1,409$

      $5 \times 1,409 = 7,045$

      Answer: 7,045 adults

20. $\$2,000 - \$370 - 4 \times \$370$

      $\$2,000 - \$370 - \$1,480$

      $= \$150$

      Answer: $150

1. C

2. B

3. C
   $28 \times 9 = 252$
   $252 + 5 = 257$

4. B

5. A

6. $(\underline{81} + \underline{19}) \times 12$

7. $(100,000 \times \underline{1,000}) - (\underline{3} \times 1,000)$

8a. $88 + 12 \div 3 - 4 \times 7$
    $= 88 + 4 - 28 = 64$

8b. $9 \times (14 + 6) \times 2$
    $= 9 \times 20 \times 2 = 360$

9. $28 - 2 \times \{[3 \times (3 + 7) + 5] \div 7\}$
   $= 28 - 2 \times \{[3 \times 10 + 5] \div 7\}$
   $= 28 - 2 \times \{[30 + 5] \div 7\}$
   $= 28 - 2 \times \{35 \div 7\}$
   $= 28 - 2 \times 5 = 28 - 10 = 18$
   Answer: 18

10. $99 \times 1,001 - 99$
    $= 99 \times (1,001 - 1)$
    $= 99 \times 1,000$
    $= 99,000$
    Answer: 99,000

11. Guess and check.
    $4 \times (5 - 3 + 6) \div 2 = 16$

12. Work backwards.
    $12,588 - 1,324 = 12,264$
    $12,264 \div 42 = 292$
    Answer: 292

13. $1,429 \times 5 = 7,145$
    Answer: 2

14. 451 R 8

15. $1,536 \times 60 = 92,160$ min

16. 5

17. After

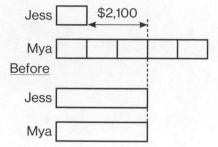

2 units $\longrightarrow$ $2,100
6 units $\longrightarrow$ $3 \times \$2,100 = \$6,300$
Answer: $6,300

18. Cost of entry for 6 children
    $= \$856 - \$508 = \$348$
    Cost of entry for 2 children
    $= \$348 \div 3 = \$116$
    Cost of entry for 4 adults
    $= \$508 - 116 = \$392$
    Cost of entry for 1 adult
    $= \$392 \div 4 = \$98$
    Answer: $98

19. $24 \div 6 \times \$32 - 24 \div 8 \times \$36$
    $= 4 \times \$32 - 3 \times \$36$
    $= \$128 - \$108$
    $= \$20$
    Answer: $20

20. After

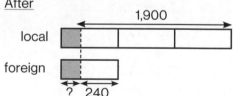

$(1,900 - 240) \div 2 - 240$
$= 1,660 \div 2 - 240$
$= 830 - 240 = 590$
Answer: 590

Unit 3 — Test A

1. C

2. A

3. C

$$\frac{1}{2} - \frac{2}{5} = \frac{5}{10} - \frac{4}{10} = \frac{1}{10}$$

4. D

$$10 \times \frac{7}{8} = 5 \times \frac{7}{4} = \frac{35}{4} = 8\frac{3}{4} \text{ yd}^2$$

5. C

$$1 - \frac{5}{6} = \frac{1}{6}$$

$$\frac{5}{6} \rightarrow 90$$

$$\frac{1}{6} \rightarrow \$90 \div 5 = \$18$$

6. $5 \times 3 = 15$
   $9 \times 3 = 27$
   Answer: 15

7. $3\frac{5}{6} - 2\frac{3}{4}$

$$= 3\frac{10}{12} - 2\frac{9}{12} = 1\frac{1}{12} = \frac{13}{12}$$

Answer: $\frac{13}{12}$

8. $1\frac{1}{2} - \frac{1}{4}$

$$= 1\frac{2}{4} - \frac{1}{4} = 1\frac{1}{4} = \frac{5}{4} = \frac{10}{8}$$

Answer: 10

9. $52 \div 6 = 8\frac{4}{6} = 8\frac{2}{3}$

Answer: $8\frac{2}{3}$

10. $\frac{3}{8} \times 9 = \frac{27}{8} = 3\frac{3}{8}$

11. $3\frac{15}{25} = 3\frac{3}{5}$

12. $\frac{11}{4}, 2\frac{1}{4}, 1\frac{5}{8}, 1$

13. $1\frac{1}{5} + 2\frac{3}{5} = 3\frac{4}{5}$

Answer: $3\frac{4}{5}$

14. $300 \div 7 = 42\frac{6}{7}$

Answer: $42\frac{6}{7}$

15. $2\frac{1}{3} - 1\frac{1}{2} = \frac{5}{6}$

Answer: $\frac{5}{6}$

16. Method 1:

$$1 - \frac{5}{8} = \frac{3}{8}$$

$$\frac{3}{8} \times 160 = 60$$

Answer: 60

Method 2:

$$\frac{5}{8} \times 160 = 100$$

$$160 - 100 = 60$$

Answer: 60

17. $2\frac{1}{4} - 1\frac{2}{3}$

$$= 2 - 1\frac{2}{3} + \frac{1}{4} = \frac{1}{3} + \frac{1}{4} = \frac{7}{12} \text{ h}$$

Answer: $\frac{7}{12}$ h

18. $1 - \frac{7}{11} = \frac{4}{11}$

$$121 \times \frac{4}{11} = 44 \text{ cm}$$

Answer: 44 cm

19. $2\frac{1}{5} + 2\frac{1}{5} + 1\frac{3}{10} = 4\frac{2}{5} + 1\frac{3}{10}$

$$= 4\frac{4}{10} + 1\frac{3}{10} = 5\frac{7}{10}$$

Answer: $5\frac{7}{10}$ m

20. $1 - \frac{1}{3} - \frac{1}{4} = \frac{5}{12}$

$$\$108 \times \frac{5}{12} = \$45$$

Answer: $45

Unit 3 — Test B

1. B
$5 \times 13 = 65$
$7 \times 13 = 91$

2. A

3. C

4. C

5. A
$46 + 2 = 48$
$\dfrac{1}{4} \times 48 = 12$
$12 - 2 = 10$

6. $1\dfrac{3}{5} + 2\dfrac{1}{3} = 1\dfrac{9}{15} + 2\dfrac{5}{15} = 3\dfrac{14}{15}$
$5 - 3\dfrac{14}{15} = 1\dfrac{1}{15}$
Answer: $1\dfrac{1}{15}$

7. $4\dfrac{3}{5} = \dfrac{23}{5}$
Answer: $\dfrac{19}{5} < 4\dfrac{3}{5}$

8. $3\dfrac{1}{9}, 3\dfrac{1}{3}, 4, \dfrac{16}{3}$

9. $\dfrac{5}{7} + \dfrac{1}{14} + \dfrac{3}{14}$
$= \dfrac{10}{14} + \dfrac{1}{14} + \dfrac{3}{14} = 1$
Answer: $\dfrac{5}{7}, \dfrac{1}{14}, \dfrac{3}{14}$

10. $6 \times \dfrac{1}{8} = 2 \times \dfrac{3}{8}$
Answer: 2

11. $12 \times 2\dfrac{1}{8}$
$= 12 \times \dfrac{17}{8}$
$= 3 \times \dfrac{17}{2}$
$= \dfrac{51}{2}$
$= 25\dfrac{1}{2}$
Answer: $25\dfrac{1}{2}$ in.^{2}

12. 2 years = 24 months
$\dfrac{30}{24} = \dfrac{10}{8} = 1\dfrac{2}{8} = 1\dfrac{1}{4}$
Answer: $1\dfrac{1}{4}$ yr

13. $\dfrac{3}{4} \longrightarrow 27$
$\dfrac{1}{4} \longrightarrow 27 \div 3 = 9$
$\dfrac{4}{4} \longrightarrow 4 \times 9 = 36$ (☆)
$\dfrac{2}{9} \times 36 = 8$
Answer: 8

14. $\dfrac{3}{5}$ of $70 = \dfrac{3}{5} \times 70 = 42$
$42 + 70 = 112$
Answer: 112 kg

15. $3\dfrac{2}{5} \times 5 = \dfrac{17}{5} \times 5 = 17$
Answer: 17 L

16. $8 \times 1\dfrac{5}{6}$
$= 8 \times \dfrac{11}{6}$
$= \dfrac{44}{3}$
$= 14\dfrac{2}{3}$
Answer: $14\dfrac{2}{3}$ L

17. $4\dfrac{1}{2} + 3\dfrac{3}{5}$
$= 4\dfrac{5}{10} + 3\dfrac{6}{10}$
$= 7\dfrac{11}{10}$
$= 8\dfrac{1}{10}$ kg
$12 - 8\dfrac{1}{10} = 3\dfrac{9}{10}$ kg
Answer: $3\dfrac{9}{10}$ kg

18. $3\frac{1}{8} \times \$2 = \$6.25$ (sand)

$3 \times \$2 = \$6$

$2\frac{1}{4} \times \$6 = \$13.50$

$\$6.25 + \$13.50 = \$19.75$
Answer: $19.75

19. $20 - 10 = 10$

$\frac{10}{3} = 3\frac{1}{3}$

Answer: $3\frac{1}{3}$ lb in each container

20. $5 \times \frac{5}{8} = \frac{25}{8} = 3\frac{1}{8}$

$3\frac{1}{8} - \frac{3}{4}$

$= 3\frac{1}{8} - 1 + \frac{1}{4}$

$= 2\frac{1}{8} + \frac{1}{4}$

$= 2\frac{3}{8}$ L

Answer: $2\frac{3}{8}$ L

Continual Assessment 1) Test A

1. B

2. B

3. B

4. B

5. C

6. D
$6 + 12 + 18 + 24 = 60$

7. D

8. B

9. D
$1 - \dfrac{2}{3} = \dfrac{1}{3}$

$\dfrac{1}{3} \times \dfrac{4}{5} = \dfrac{4}{15}$

10. D

$\dfrac{5}{6} \longrightarrow 30$

$\dfrac{1}{6} \longrightarrow 30 \div 5 = 6$

$\dfrac{6}{6} \longrightarrow 6 \times 6 = 36$

$\dfrac{1}{3} \times 36 = 12$

11. $500{,}000{,}000 + 5{,}000{,}000 + 50{,}000 + 5{,}000 + 50 + 5$

12. Count on by 120,000s.
744,307

13. 999,999

14. 314,498

15. $10^5 = 10 \times 10 \times 10 \times 10 \times 10 = 100{,}000$

16. 44

17. 78

18. $2 = \dfrac{10}{5}$

$10 - 4 = 6$
Answer: 6

19. 47 R 12

20. 45

21. $5{,}280 \div 60 = 88$
$88 - 1 = 87$
Answer: 87 cuts

22. $200 \times \$7 + (600 - 200) \div 2 \times \$11 - \$2{,}400$
$= \$1{,}400 + 400 \div 2 \times \$11 - \$2{,}400$
$= \$1{,}400 + 200 \times \$11 - \$2{,}400$
$= \$1{,}400 + \$2{,}200 - \$2{,}400$
$= \$1{,}200$
Answer: \$1,200

23. $3{,}954 \times 2 = 7{,}908$
$3{,}954 + 7{,}908 + 1{,}498 = 13{,}360$
Answer: 13,360 people

24. $9\dfrac{1}{4} + 2\dfrac{3}{4} = 12$ (length)

$12 \times 9\dfrac{1}{4} = 111$

Answer: $111\,\text{m}^2$

25. $(10 - 7)$ units -> 30
1 unit $\longrightarrow$ 10
10 units $\longrightarrow$ 100
Answer: 100 chickens

Extra Credit

1. X's thousands digit: 4
X's hundreds digit: $2 \times 4 = 8$
X stands for 4,800
$Y = 8{,}400 - 4{,}800 = 3{,}600$
Answer: 3,600

2.   Method 1:
Use replacement concept.
Let $\bigcirc$ be the cost of a shirt and $\square$ be
the cost of a tie.
$\bigcirc = \square + \$40$
$2\bigcirc = 2\square + \$80$
$2\bigcirc + 3\square = (2\square + \$80) + 3\square$
$\quad = 5\square + \$80$
$5\square + \$80 = \$230$
$5\square = \$230 - \$80 = \$150$
$\square = \$150 \div 5 = \$30$
$\bigcirc = \$30 + \$40 = \$70$

Method 2:
Total cost of 5 shirts and 5 ties
$= (\$230 + \$40) + \$230 = \$500$
Cost of 1 shirt and 1 tie
$= \$500 \div 5 = \$100$
Cost of 1 shirt
$= \$270 - (\$100 \times 2)$
$= \$270 - \$200$
$= \$70$
Answer: $70

1. D

2. B
$(682 - 60) \div 2 = 311$

3. A

4. C
$340 - 130 + 90 = 300$

5. A

6. B

7. A

8. C
$$\frac{48}{10} = 4\frac{8}{10} = 4\frac{4}{5}$$

9. D
$$\frac{9}{4} = 2\frac{1}{4} = 2\frac{2}{8}$$

10. A

11. 8

12. 1,600,000

13. 3,080,898

14. 21 R 26

15. $2 \times 2 \times 5 \times 5 = 2^2 \times 5^2$

16. $22 \times 70 + 50 = 1,590$

17. $5 \times 6 + 12 \div 3 \times (4 + 5)$
$= 30 + 4 \times 9$
$= 30 + 36$
$= 66$

18. $\frac{9}{10} - \frac{2}{5}$
$= \frac{9}{10} - \frac{4}{10} = \frac{5}{10} = \frac{1}{2}$
Answer: $\frac{1}{2}$

19. $\frac{2}{3} + \frac{1}{6} + \frac{7}{12}$
$= \frac{8}{12} + \frac{2}{12} + \frac{7}{12} = \frac{17}{12} = 1\frac{5}{12}$
Answer: $1\frac{5}{12}$

20. 156 minutes

21. $15 \times 28 \div 20 = 21$
Answer: 21 rows

22. $180 - 156 = 24$
$\$1,080 - \$600 = \$480$
$\$480 \div 24 = \$20$
Answer: $20

23.
$18 + $27

5 units → $18 + $27 = $45
1 unit → $45 ÷ 5 = $9
7 units → 7 × $9 = $63
Answer: $63

24. $8 \times 4\frac{1}{6} = 8 \times \frac{25}{6} = \frac{100}{3}$ m^2

$\frac{100}{3} \times \$27 = \$900$

Answer: $900

25. $1 - \frac{2}{7} = \frac{5}{7}$

$\frac{5}{7}$ → $102 - 37 = 65$

$\frac{1}{7}$ → $65 \div 5 = 13$

$\frac{7}{7}$ → $7 \times 13 = 91$

$102 - 91 = 11$ kg
Answer: 11 kg

Extra Credit

1. ($900 – 30 × $8) ÷ ($12 + $8) + 30
    = ($900 – $240) ÷ $20 + $30
    = $660 ÷ $20 + 30
    = 33 + 30 = 63
    Answer: 63 children

2.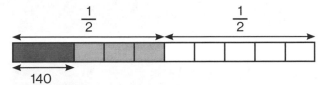

    $\frac{1}{2}$ $\frac{1}{2}$

    140

    2 units = 140
    1 unit = 140 ÷ 2 = 70
    10 units = 70 × 10 = 700
    Answer: 700 tarts

Unit 4 Test A

1. B

2. D

$$\frac{3}{4} \div 4 = \frac{3}{4} \times \frac{1}{4} = \frac{3}{16}$$

3. C

$$\frac{1}{3} \div 3 = \frac{1}{3} \times \frac{1}{3} = \frac{1}{9}$$

4. A

$$\frac{5}{6} \div \frac{5}{12} = \frac{5}{6} \times \frac{12}{5} = 2$$

5. A

6. $\frac{3}{4} \times \frac{5}{6} = \frac{5}{8}$

7. $\frac{3}{5} \times \frac{10}{11} = \frac{6}{11}$

8. $1\frac{3}{5} \times 10 = \frac{8}{5} \times 10 = 16$

9. $\frac{11}{2} \times \frac{8}{5} = \frac{11}{1} \times \frac{4}{5} =$

$\frac{44}{5} = 8\frac{4}{5}$

Answer: $8\frac{4}{5}$

10. $3 \div \frac{1}{2} = 3 \times 2 = 6$

11. $8 \div \frac{4}{5} = 8 \times \frac{5}{4} = 10$

12. $\frac{3}{4} \div \frac{1}{8} = \frac{3}{4} \times 8 = 6$

13. $\frac{3}{5} \div \frac{3}{10} = \frac{3}{5} \times \frac{10}{3} = 2$

14. $\frac{2}{5} \div 4 = \frac{1}{10}$ m

Answer: $\frac{1}{10}$ m

15. $10 \div \frac{3}{4} = 10 \times \frac{4}{3} = \frac{40}{3} = 13\frac{1}{3}$

Answer: The length is $13\frac{1}{3}$ ft

16. $\frac{3}{4} \times \frac{4}{5} = \frac{3}{5}$ L

Answer: $\frac{3}{5}$ L

17. $\frac{3}{10} \times (1 - \frac{1}{6}) = \frac{1}{4}$

Answer: $\frac{1}{4}$

18. $\frac{2}{3} \div 4 = \frac{2}{3} \times \frac{1}{4} = \frac{1}{6}$

Answer: Each friend gets $\frac{1}{6}$ of the pizza

19. $\frac{9}{10} \div 3 = \frac{3}{10}$ kg

Answer: $\frac{3}{10}$ kg

20. $2\frac{1}{3} \times 5\frac{1}{2} = \frac{7}{3} \times \frac{11}{2} = \frac{77}{6} = 12\frac{5}{6}$

Answer: $12\frac{5}{6}$ L

1.   C

2.   B

$$\frac{3}{8} \times 2 = \frac{3}{4} \, m$$

$$\frac{3}{8} \times \frac{3}{4} = \frac{9}{32} \, m^2$$

3.   D

4.   D

$$(\frac{1}{4} + \frac{1}{6}) \div 2 = \frac{5}{24}$$

5.   A

$$\frac{3}{4} - \frac{1}{8} = \frac{5}{8}$$

5 units $\longrightarrow$ 30 L

1 unit $\longrightarrow$ 30 ÷ 5 = 6 L

8 units $\longrightarrow$ 8 × 6 = 48 L

Answer: 48 L

6. $\frac{3}{7} \div 12 = \frac{3}{7} \times \frac{1}{12} = \frac{1}{7} \times \frac{1}{4} = \frac{1}{28}$

Answer: $\frac{1}{28}$

7. $9 \div \frac{3}{4} = 9 \times \frac{4}{3} = 12$

8. $20 \div \frac{5}{9} = 20 \times \frac{9}{5} = 36$

9. $\frac{2}{3} \div \frac{4}{9} = \frac{2}{3} \times \frac{9}{4} = \frac{3}{2} = 1\frac{1}{2}$

10. $\frac{5}{6} \div \frac{5}{12} = \frac{5}{6} \times \frac{12}{5} = 2$

11. $\frac{2}{3} \div \frac{4}{9} = \frac{2}{3} \times \frac{9}{4} = \frac{1}{1} \times \frac{3}{2} = 1\frac{1}{2}$

Answer: The other number is $1\frac{1}{2}$

12. $\frac{4}{5} \div \frac{1}{10} = \frac{4}{5} \times 10 = 8$

Answer: 8

13. $3\frac{1}{2}$ of $1\frac{4}{7} = \frac{7}{2} \times \frac{11}{7} = \frac{11}{2} = 5\frac{1}{2}$

Answer: $5\frac{1}{2}$ cups

14. $\frac{2}{3} \div \frac{4}{9} = \frac{2}{3} \times \frac{9}{4} = \frac{3}{2} = 1\frac{1}{2} \, m$

Answer: $1\frac{1}{2} \, m$

15. $8 \div \frac{2}{3} = 8 \times \frac{3}{2} = 12$

$\frac{1}{2}$ of $12 = \frac{1}{2} \times 12 = 6$

$\frac{1}{2}$ sack of rice = 6 lb

16. $1\frac{5}{9} = \frac{14}{9}$

$\frac{14}{9} \div 7 = \frac{14}{9} \times \frac{1}{7} = \frac{2}{9}$

Answer: Each part is $\frac{2}{9}$ yd

17. $\frac{1}{3} \times \frac{4}{7} = \frac{4}{21}$

$\frac{4}{21} \longrightarrow 264$

$\frac{1}{21} \longrightarrow 264 \div 4 = 66$

$\frac{21}{21} \longrightarrow 21 \times 66 = 1{,}386$

Answer: 1,386 students

18. $\frac{3}{4} \times (1 - \frac{1}{3}) = \frac{1}{2}$

$1 - \frac{1}{3} - \frac{1}{2} = \frac{1}{6}$

$\frac{1}{6} \longrightarrow 75$

$\frac{6}{6} \longrightarrow 6 \times 75 = 450$

Answer: 450 pages

19. $\frac{2}{5} \times \frac{3}{8} = \frac{3}{20}$

3 units $\longrightarrow$ 147

1 unit $\longrightarrow$ 49

20 units $\longrightarrow$ 980 balls

20. $1 - \frac{1}{3} = \frac{2}{3}$

$\frac{5}{6} \times \frac{2}{3} = \frac{5}{9}$

$1 - \frac{1}{3} - \frac{5}{9} = \frac{1}{9}$

Answer: $\frac{1}{9}$

Unit 5 — Test A

1. B
$3 \times 2 = 6$ cm
$4 \times 2 = 8$ cm
$\frac{1}{2} \times 6 \times 8 = 24$ cm^2

2. B

3. C

4. A
$\frac{1}{2} \times 6 \times 6 = 18$ cm^2

5. B
$7 \times 2\frac{1}{5} = 7 \times \frac{11}{5} = \frac{77}{5} = 15\frac{2}{5}$

6. $\frac{1}{2} \times 3 \times 4 = 6$ cm^2
Answer: 6 cm^2

7. $6 + 9 = 15$
Perimeter = $15 \times 2 = 30$ cm
$9 - 6 = 3$
$6 \times 6 = 36$
$3 \times 3 = 9$
Area = $36 + 9 = 45$ cm^2

8. $3 \times 12 = 36$
$36 + 72 = 108$
$2 \times 108 = 216$
Perimeter = $216 + 9 + 9 = 234$ m
$72 \times 36 = 2,592$
$9 \times 12 = 108$
Area = $2,592 - 108 = 2,484$ m^2

9. B

10. $\frac{1}{2} \times 15\frac{1}{3} \times 8 = \frac{1}{2} \times \frac{46}{3} \times 8 = \frac{46}{3} \times 4$
$= \frac{184}{3} = 61\frac{1}{3}$ cm^2
Answer: $36\frac{4}{5}$ cm^2

11. $\frac{1}{2} \times 12 \times 9 = 54$ cm^2
Answer: 54 cm^2

12. Draw a diagram.

$112 \div 8 = 14$
$3 \times 14 = 42$
Answer: 42 m

13. Area of rectangle = $20 \times 10 = 200$
Base of triangle = $10 - 4 = 6$
Area of triangle = $\frac{1}{2} \times 6 \times 20 = 60$
Area of shaded part = $200 - 60 = 140$
Answer: 140 in.^{2}

14. $20 + 12 = 32$
Perimeter = $20 + 20 + 32 + 32 = 104$ m
Answer: 104 m

15. Area of triangle = $\frac{1}{2} \times (9 \times 8) = 36$
Area of parallelogram = $9 \times 5 = 45$
Area of figure = $36 + 45 = 81$ ft^2

16. Area of unshaded parts = area of rectangle − area of shaded triangle
Area of shaded triangle
$= \frac{1}{2}$ of rectangle
So, area of unshaded parts
$= \frac{1}{2}$ of rectangle
$= \frac{1}{2} \times 23 \times 28$
$= 322$ cm^2
Answer: 322 cm^2

17. $30 - 16 = 14$ cm
Area of shaded part
$= \frac{1}{2} \times 14 \times 16$
$= 112$ cm^2
Answer: 112 cm^2

18. $27 + 21 = 48$ cm
Area of shaded part
$= \frac{1}{2} \times 48 \times 27 = 648$ cm^2
Answer: 648 cm^2

19. $24 \times 10 = 240$ m^2
$6 \times 6 = 36$ m^2
$240 - 36 = 204$ m^2
Answer: 204 m^2

20. $52 \div 2 = 26$

$(26 - 8) \div 2 = 9$ (width)
$9 + 8 = 17$ (length)
$17 \times 9 = 153$
Answer: 153 m^2

Unit 5   Test B

1. D
    $6 \times 6 = 36 \text{ cm}^2$

    $\frac{1}{2} \times 6 \times 4 = 12 \text{ cm}^2$

    $\frac{1}{2} \times 2 \times 3 = 3 \text{ cm}^2$

    $\frac{1}{2} \times 6 \times 3 = 9 \text{ cm}^2$

    $36 - 12 - 3 - 9 = 12 \text{ cm}^2$

2. C
    $\frac{1}{2} \times 6 \times (8 + 3) = 33 \text{ cm}^2$

    $\frac{1}{2} \times 3 \times 6 = 9 \text{ cm}^2$

    $33 + 9 = 42 \text{ cm}^2$

3. B
    Restate the problem.
    Draw a surrounding square with 12-cm
    sides as shown below.

    Area of square ABCD

    $= \frac{1}{2}$ of surrounding square

    $12 \times 12 \div 2 = 72 \text{ cm}^2$

4. D
    $12 \div 2 = 6 \text{ cm}$
    $10 \times 6 + 8 \times 6 = 108 \text{ cm}^2$

    $\frac{1}{2} \times 10 \times 12 = 60 \text{ cm}^2$

    $\frac{1}{2} \times 6 \times 2 = 6 \text{ cm}^2$

    $\frac{1}{2} \times 6 \times 8 = 24 \text{ cm}^2$

    $108 - 60 - 6 - 24 = 18 \text{ cm}^2$

5. A
    Base of each shaded triangle = 6 cm
    Total height of two shaded triangles =
    10 cm
    Total area of two shaded triangles

    $= \frac{1}{2} \times 6 \times 10 = 30 \text{ cm}^2$

    Answer: 30 cm^2

6. $\frac{1}{2} \times 4 \times 2 = 4$

    $\frac{1}{2} \times 2 \times 2 = 2$

    $4 + 2 = 6$
    Answer: 6 cm^2

7. BCEF or ABCD

8. $2 \times 4 = 8 \text{ cm}$

    $2 \times \frac{1}{2} \times 4 \times 8 = 32 \text{ cm}^2$

    Answer: 32 cm^2

9. Area of small parallelogram =

    $1 \times \frac{3}{4} = \frac{3}{4}$

    Area of big parallelogram =

    $1\frac{1}{8} \times 2 = \frac{9}{8} \times 2 = \frac{9}{4} = 2\frac{1}{4}$

    Area of figure $= \frac{3}{4} + 2\frac{1}{4} = 3 \text{ in.}^2$
    Answer: 3 in.^{2}

10. Area of CDG $= \frac{1}{2}$ of Area of ABCD

    Area of EFG $= \frac{1}{3}$ of Area of CDG

    $\frac{1}{2} \times \frac{1}{3} = \frac{1}{6}$

    Total area of shaded parts

    $= (1 - \frac{1}{2} - \frac{1}{6}) \times 108 = 36 \text{ cm}^2$

    Answer: 36 cm^2

11. $9 + 9 = 18$
    $18 + 40 = 58$
    $58 \times 2 = 116$
    Perimeter $= 116 + 9 + 9 = 134 \text{ cm}$
    $40 \times 9 = 360$
    $9 \times 9 = 81$
    Area $= 360 + 81 + 81 = 522 \text{ cm}^2$
    Answer: 522 cm^2

12. $18 \div 3 = 6$

    $\frac{1}{2} \times 6 \times 26 = 78 \text{ cm}^2$

    Answer: 78 cm^2

13. $72 \div 4 = 18 \text{ cm}$

    $3 \times \frac{1}{2} \times 18 \times 18 = 486 \text{ cm}^2$

    Answer: 486 cm^2

14. $112 \div 4 = 28$
$28 \div 2 = 14$
$14 \times 28 = 392$
Answer: 392 cm^2

15. The perimeter of the remaining piece of paper is the same as the original square piece of paper.
$9 \times 9 = 81$
So, length = 9 cm
Perimeter = $9 \times 4 = 36$ cm
Answer: 36 cm

16. $\frac{1}{2} \times 8 \times 6\frac{1}{3} = 4 \times \frac{19}{3} = \frac{76}{3} = 25\frac{1}{3} \text{ cm}^2$

$\frac{1}{2} \times 10 \times 4 = 20 \text{ cm}^2$

$25\frac{1}{3} - 20 = 5\frac{1}{3} \text{ cm}^2$

Answer: $5\frac{1}{3} \text{ cm}^2$

17. $1{,}000 \div 4 = 250$
$250 \div 2 = 125$
$125 - 35 = 90$
Answer: 90 m

18. Area of wall = $900 = 30 \times 30$
Height of wall = 30
Half height of wall = $30 \div 2 = 15$
Half width of painting = $7 \div 2 = 3\frac{1}{2}$

Ground to bottom edge of

painting = $15 - 3\frac{1}{2} = 11\frac{1}{2}$ ft

19. $\frac{1}{2} \times 12 \times 9 = 54 \text{ cm}^2$

$4 \times 54 = 216 \text{ cm}^2$
$(12 - 9) \times (12 - 9) = 9 \text{ cm}^2$
$216 + 9 = 225 \text{ cm}^2$
Answer: 225 cm^2

20. $15 \times 15 = 225$
Answer: 15 cm

1. A
    $10:15 = 2:3$

2. A
    $8:(8-2) = 8:6 = 4:3$

3. D

4. B
    3 units ⟶ 18 liters
    1 unit ⟶ $18 \div 3 = 6$ liters
    2 units ⟶ $2 \times 6 = 12$ liters

5. C
    $(5-3)$ units ⟶ 10 kg
    $(5+3)$ units ⟶ $4 \times 10 = 40$ kg

6. $4:3$

7a. $7:3$

7b. $3:7$

7c. $10:7$

8a. $2:1$

8b. $1:3$

9. $3:4:6$

10. $60:40:30$ or $6:4:3$

11a. $\underline{6}:8:\underline{2}$

11b. $\underline{20}:\underline{16}:12$

12a. $4:5:2$

12b. $6:3:10$

13. Divide the stars into 3 equal groups,
    then shade 2 such groups.

    $6-2 = 4$
    Answer: 4 more stars

14. $12 + 6 = 18$
    $12:18 = 2:3$
    Answer: $2:3$

15. $5:1$

16. $\$25 + \$15 = \$40$
    $25:40 = 5:8$
    Answer: $5:8$

17a. Week 2 campers = $40 - 24 = 16$
    $24:16 = 3:2$
    Answer: $3:2$

17b. $16:40 = 2:5$
    Answer: $2:5$

18a. $16:9$

18b. $12:8 = 3:2$
    Answer: $3:2$

18c. $9:12 = 3:4$
    Answer: $3:4$

18d. $8:45$

19. 2×60 min $= 120$ min
    $15:120 = 1:8$
    Answer: $1:8$

20. $\$30 \div 6 = \$5$
    $(6+4+5) \times \$5 = \$75$
    Answer: $\$75$

1.   D

    5 units $\longrightarrow$ 1 + 3 + 5 + 7 + 9 = 25

    1 unit $\longrightarrow$ 25 ÷ 5 = 5

    3 units $\longrightarrow$ 3 × 5 = 15

    15 − 9 = 6

2.   B

3.   C

    R : Y = 2 : 3 = 4 : 6

    Y : B = 2 : 3 = 6 : 9

    So, R : B = 4 : 9

4.   B

    2 × 12 = 24

    (12 + 8) : (24 + 8) = 20 : 32 = 5 : 8

5.   D

6a.   6 : 4 = 3 : 2

    Answer: 3 : 2

6b.   4 : 6 = 2 : 3

    Answer: 2 : 3

6c.   4 : 10 = 2 : 5

    Answer: 2 : 5

7.   4 : 8 : 6 = 2 : 4 : 3

    Answer: 2 : 4 : 3

8a.   11 : 5 : 3 = <u>55</u> : 25 : <u>15</u>

8b.   10 : 6 : 9 = <u>60</u> : <u>36</u> : 54

9a.   5 : 1 : 3

9b.   3 : 2 : 9

9c.   6 : 4 : 5

9d.   3 : 4 : 1

10a.   4 : 3

10b.   6 : 5

10c.   7 : 25

10d.   25 : 4

11.   Perimeter of A = 9 + 12 + 15 = 36 cm

    Perimeter of B = 3 × 15 = 45 cm

    Perimeter of C = 2 × 22 + 10 = 54 cm

    36 : 45 : 54 = 4 : 5 : 6

    Answer: 4 : 5 : 6

12.   3 : 1 : 2

13.   Area of A = 12 × 4 = 48 cm^2

    Area of B = 8 × 8 = 64 cm^2

    48 : 64 = 3 : 4

    Answer: 3 : 4

14.   Base of triangle = 8 ft

    Area of triangle = $\frac{1}{2}$ × 8 × 12 = 48 ft^2

    Answer: 48 ft^2

15.   84 ÷ 2 × 5 = 210 cm

    Answer: 210 cm

16.   30 ÷ 5 × 9 = 54

    Answer: 54 papayas

17.   30 ÷ 5 × 2 = 12

    Answer: 12 liters

18a.   3 units = 36

    1 unit = 36 ÷ 3 = 12

    7 units = 7 × 12 = 84

    Answer: 84

18b.   7 + 3 = 10

    10 units = 10 × 12 = 120

    Answer: 120

19.   (4 + 5 + 3) units $\longrightarrow$ 60 cm

    1 unit $\longrightarrow$ 60 ÷ 12 = 5 cm

    5 units $\longrightarrow$ 5 × 5 = 25 cm

    Answer: 25 cm

20.   Before

    R : L = 2 : 5 = 6 : 15

    After

    8 − 6 = 2 units

    2 units $\longrightarrow$ 30

    1 unit $\longrightarrow$ 30 ÷ 2 = 15

    (6 + 15) units $\longrightarrow$ 21 × 15 = 315

    Answer: 315 roses and lilies

1. D

2. C

3. C

4. B
   ($330 − $80) ÷ 5 = $50

5. A

6. B

7. C

8. C

   $$3\frac{1}{4} = 3\frac{3}{12}$$

   $$2\frac{5}{12} + 3\frac{3}{12} = 5\frac{8}{12} = 5\frac{2}{3}$$

9. A

10. B

11. A
    Use restate-the-problem strategy.
    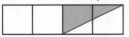

12. A

13. D

14. D

15. A
    9 units ⟶ 36
    1 unit ⟶ 4
    4 units ⟶ 16

16. 3,035,250

17. 5,000

18. $6^3 = 6 \times 6 \times 6 = 216$

19. 3,000

20. $67,000

21. 1,500,000

22. 10, 16, 20

23. $35 − 5 \times (42 ÷ 7) = 5$
    $5 \times 100 = 500$

Extra Credit

1.   Before

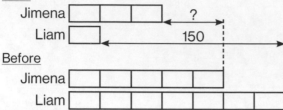

6 units = $150
1 unit = $150 ÷ 6 = $25
2 units = 2 × $25 = $50
Answer: $50

2.   The number of paper cranes Sun-mi folded
       remained the same.
       <u>Before</u>
           Noelani : Sun-mi = 3 : 7 = 9 : 21
       <u>After</u>
           Noelani : Sun-mi = 2 : 3 = 14 : 21
       14 − 9 = 5
       5 units = 35
       1 unit = 35 ÷ 5 = 7
       14 units = 14 × 7 = 98
       Answer: 98

1. B

2. D

3. C

4. C
When 2,998 is rounded to the nearest 10, 100 or 1,000, the answer is 3,000.

5. C
$2 + 4 = 6$

6. A

7. A

8. A

9. A

10. D

11. A

12. B
2 units ⟶ $44
1 unit ⟶ $22
5 units ⟶ $110
$110 + $60 = $170

13. A

14. D
$(\$96 - \$16) \div 2 = \$40$
$\$40 + \$16 = \$56$
$\$40 : \$56 = 5 : 7$

15. A

16. $100,000,000 + 80,000,000 + 300,000 + 50,000 + 100 + 20 + 3$

17. 100

18. millions
90,000
0

19. Multiples of 3: 3, 6, 9, <u>12</u>
Mulitples of 6: 6, <u>12</u>
Multiples of 12: <u>12</u>
Answer: 12

20. Factors of 30: 1, 2, 3, 5, 6, <u>10</u>, 15, 30
Factors of 50: 1, 2, 5, <u>10</u>, 25, 50
Answer: 10

21. $5^3 = 5 \times 5 \times 5 = 125$
$3^5 = 3 \times 3 \times 3 \times 3 \times 3 = 243$
Answer: $5^3 < 3^5$

22.

2	800
2	400
2	200
2	100
2	50
5	25
5	5
	1

$2 \times 2 \times 2 \times 2 \times 2 \times 5 \times 5 = 2^5 \times 5^2$
Answer: $2^5 \times 5^2$

23. $73 \times 46 = 3,358$
$3,358 + 20 = 3,378$
Answer: 3,378

24. $33 - (3 + 4) \times 3 - 10 \div 5$
$= 33 - 7 \times 3 - 2 = 33 - 21 - 2 = 10$
Answer: 10

25. $(\frac{2}{3} - \frac{1}{4}) = \frac{5}{12}$
$\frac{5}{12} \longrightarrow 760$
$\frac{12}{12} \longrightarrow 760 \div \frac{5}{12} = 1,824$
Answer: 1,824 ml

26. $23 + 19 - 26 = 16$
Answer: 16 cm

27. The length of the wire used to form the figure is 5 times as long as PQ.
$5 \times 37 = 185$ cm
$300 - 185 = 115$ cm
Answer: 115 cm

28. $12 \times 12 = 144$ and $18 \times 18 = 324$
$\frac{1}{2} \times 12 \times (18 - 12) = 36$ cm^2

29a. $(44 - 2 \times 10) \div 6 = 4$ cm

29b. $4 \times 4 + \frac{1}{2} \times 4 \times 4 = 24$
$(2 \times 4 + 10) \times 4 = 72$
$\frac{24}{72} = \frac{1}{3}$
Answer: $\frac{1}{3}$

30a.   Length of longer piece
= 20 − 8 = 12
12 : 8 = 3 : 2
Answer: 3 : 2

30b.   20 : 12 = 5 : 3
Answer: 5 : 3

31.   (3,600 − 390) ÷ 5 = 642 (C)
642 × 2 + 390 = 1,674 books

32.   Use draw-a-model strategy.

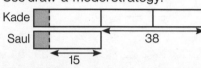

38 ÷ 2 = 19
19 − 15 = 4 years

33. $\dfrac{2}{5} \times \dfrac{5}{6} = \dfrac{1}{3}$

1 unit ⟶ $380
3 units ⟶ $1,140

34. $\dfrac{1}{2} \times (18 − 14) \times (18 − 10) = 16 \text{ cm}^2$

18 × 18 = 324 cm^2
324 − 16 = 308 cm^2
Answer: 308 cm^2

35.   Use before-after concept.
Before:
3 : 2 = 15 : 10
15 units ⟶ 300
1 unit ⟶ 20
After:
2 : 5 = 4 : 10
(15 − 4) units ⟶ 11 × 20 = 220

Extra Credit

1.   Area of triangle ADJ
= $\dfrac{1}{2}$ of area of rectangle ABCD
= $\dfrac{1}{2} \times 120$
= 60 cm^2

AE = HD = $\dfrac{1}{5}$ of AD

Area of triangle AEJ or HDJ
= $\dfrac{1}{5} \times 60 = 12$ cm^2
2 × 12 = 24 cm^2
Answer: 24 cm^2

2.   The number of green marbles
remained the same.
Before
   Red : Green : Blue = 7 : 3 : 12
After
   Red : Green = 3 : 1 = 9 : 3
9 − 7 = 2
2 units = 24
1 unit = 24 ÷ 2 = 12
3 units = 3 × 12 = 36
Answer: 36